THE COSMIC REVOLUTIONARY'S HANDBOOK
(Or: How to Beat the Big Bang)

For all those who wonder about our bizarre universe, and those who might want to overthrow the Big Bang with alternative theories, this handbook gives you "just the facts": the observations that have shaped these ideas and theories. While the Big Bang holds the attention of scientists, it isn't perfect. The authors pull back the curtains, and show how cosmology really works, by asking "If you wanted to replace the Big Bang model, how would you go about it? What cosmological evidence would you need to account for?" Only by accounting for these observations – all of them – can you present a viable cosmological theory.

This uniquely-framed tour of modern cosmology gives a deep understanding of the inner workings of this fascinating field. The portrait painted is realistic and raw, not idealized and airbrushed – it is science in all its messy detail, which doesn't pretend to have all the answers.

Luke A. Barnes is a lecturer in physics at Western Sydney University, with a Ph.D. in astronomy from the University of Cambridge. The focus of his research has been the cosmic evolution of matter, and he has published papers in the field of galaxy formation and evolution, and on the fine-tuning of the universe for life. He returned to the University of Sydney in 2011 as a Super Science Fellow, before being awarded a prestigious Templeton Fellowship to expand his research on fine-tuning of the laws of physics for complexity and life. Dr Barnes is an accomplished speaker to professional and amateur audiences, and can speak across the boundaries of cosmology, philosophy, and religion. He has lectured to numerous amateur astronomical groups and to public audiences, including speaking on fine-tuning at the Royal Institution in London in 2017. He tweets @lukebarnesastro.

Geraint F. Lewis is Professor of Astrophysics at the Sydney Institute for Astronomy, part of the University of Sydney's School of Physics. The focus of his research is cosmology and the dark side of the universe, namely the dark matter and dark energy that dominate cosmological evolution. He has published more than three hundred academic papers and is an acclaimed teacher. He also has a significant outreach profile, writing regularly for *New Scientist* and *The Conversation*, as well as regularly speaking publically on all aspects of cosmology and astronomy, including speaking at the Royal Institution in London. He also has extensive experience of interactions with the media, including podcast, radio, and television. He currently is Deputy Director of the Sydney Informatics Hub, developing the infrastructure and knowledge base to support big data, informatics, deep learning, and artificial intelligence at the University of Sydney. He tweets @Cosmic_Horizons.

"Overthrowing all of modern cosmology isn't easy, but it could happen. Maybe you will be the one to do it! If you're up for the challenge, Luke A. Barnes and Geraint F. Lewis tell you exactly what you have to accomplish. Even if you don't topple the stodgy edifice of modern science, you'll certainly learn some exciting things about the universe along the way."

Sean Carroll, author of *Something Deeply Hidden: Quantum Worlds and the Emergence of Spacetime*

"If you are looking for a fun rendezvous with the universe, this is the book for you! Barnes and Lewis help you understand the basics of cosmology with simplicity and clarity – quite a feat given the complexity of our universe."

Priyamvada Natarajan, author of *Mapping the Heavens: The Radical Scientific Ideas that Reveal the Cosmos*

"Are you unhappy with the state of cosmology and think it needs to be revolutionised? If so, cosmologists Luke Barnes and Geraint Lewis have written The Cosmic Revolutionary's Handbook *just for you. [This is] a great starting point for budding astronomers or cosmologists who want to be able answer the 'but how do we know …?' questions they're likely to get asked. 4/5 stars"

Chris North, BBC *Sky at Night Magazine*

"In their book, the authors describe the unbiased observations of the Universe that any cosmological theory has to explain. If you already have a basic background knowledge of cosmology and want to learn more about its intricacies (and to argue successfully with the revolutionaries), then this is definitely the book for you."

Keith Cooper, *Astronomy Now*

"Luke Barnes and Geraint Lewis set out to describe why cosmologists believe in the Hot Big Bang model and all the apparent complications it brings. [In addition] the authors provide a remarkably accurate insight into how science really is done. I feel that many professional scientists would benefit from reading this book! Overall, [this] is a must-read for anyone interested in better understanding why cosmologists believe all those very strange things about the Universe."

Sunny Vagnozzi, *Nature Astronomy*

"*Barnes and Lewis inform the general reader about many fascinating aspects of astronomy, astrophysics, and cosmology. The book is full of scientific facts and clarifying figures. More importantly, it clarifies the routes that led to major scientific results. … [It] will inform and educate those who respect science and are willing to learn about good science and how it is done. This should be required reading for all college students, regardless of their major.*"

V. V. Raman, *Choice Reviews*

"*This book is a popular account of modern cosmology, with more emphasis than most similar books on two important aspects: how conclusions are arrived at and which conclusions depend on which observations. I enjoyed reading this book; it's a breezy but careful introduction to where we are in our understanding of the Universe and how we got there.*"

Phillip Helbig, *The Observatory*

The Cosmic Revolutionary's Handbook

(Or: How to Beat the Big Bang)

LUKE A. BARNES
Western Sydney University

GERAINT F. LEWIS
Sydney Institute for Astronomy

CAMBRIDGE
UNIVERSITY PRESS

CAMBRIDGE
UNIVERSITY PRESS

University Printing House, Cambridge CB2 8BS, United Kingdom

One Liberty Plaza, 20th Floor, New York, NY 10006, USA

477 Williamstown Road, Port Melbourne, VIC 3207, Australia

314–321, 3rd Floor, Plot 3, Splendor Forum, Jasola District Centre, New Delhi – 110025, India

79 Anson Road, #06–04/06, Singapore 079906

Cambridge University Press is part of the University of Cambridge.

It furthers the University's mission by disseminating knowledge in the pursuit of education, learning, and research at the highest international levels of excellence.

www.cambridge.org
Information on this title: www.cambridge.org/9781108486705
DOI: 10.1017/9781108762090

Hardback first published 2020
Paperback first published 2023

Printed in the United Kingdom by TJ International Ltd, Padstow Cornwall

A catalogue record for this publication is available from the British Library.

Library of Congress Cataloging-in-Publication Data
Names: Barnes, Luke A., 1983– author. | Lewis, Geraint F., author.
Title: The cosmic revolutionary's handbook : (or: how to beat the big bang) / Luke A. Barnes, Western Sydney University, Geraint F. Lewis, Sydney Institute for Astronomy.
Description: Cambridge, United Kingdom ; New York, NY, USA : Cambridge University Press, 2020. | Includes bibliographical references and index.
Identifiers: LCCN 2019037637 (print) | LCCN 2019037638 (ebook) | ISBN 9781108486705 (hardback) | ISBN 9781108762090 (epub)
Subjects: LCSH: Cosmology–Popular works.
Classification: LCC QB982 .B386 2020 (print) | LCC QB982 (ebook) | DDC 523.1–dc23
LC record available at https://lccn.loc.gov/2019037637
LC ebook record available at https://lccn.loc.gov/2019037638

ISBN 978-1-108-48670-5 Hardback
ISBN 978-1-009-24578-4 Paperback

Luke
For Bernadette,
who will understandably prefer Terry
Pratchett's version of the big bang theory.

Geraint
To slightly misquote *Rocky Horror*,
this book is for those that rose tint my world
and keep me safe from my trouble and pain.

CONTENTS

PREFACE

This book was born from experience. We are cosmologists: our day job is to unravel the inner workings of our universe (and, occasionally, a few other universes). We have a deep love of our field and have taken many opportunities to spread the message of scientific enquiry beyond our less-than-ivory towers of academia. Writing and speaking, we have explained the view through the latest generation of telescopes, and the astounding and/or confusing theories that describe what is seen.

We genuinely enjoy interacting with the public, especially the barrage of questions that inevitably follow a talk about dark matter or the cosmic microwave background or life in the cosmos. Forced to think on our feet, the simplest questions sometimes need the most intricate answers. Most of all, we have learnt to never underestimate curious children!

We have also met others, either in person, in neatly written letters, or via grammar-optional emails, who don't like what we have said. To them, the universe we portray, the universe described by modern cosmology, just does not seem right. They don't like the reliance on unseen things ("Dark matter? You must be kidding!"). Or they distrust the bendy and stretchy space and time of Einstein's general theory of relativity. Or the weirdness of quantum mechanics. Or the conclusion that there must have been a cosmic birth. Or the thought of a bleak and frigid future ahead. Clearly modern cosmology has been led astray, away from logic and common sense, to something complex and crazy.

Some of these people have been working hard to fix this problem, developing their own ideas about the real universe, based on solid logic, that will revolutionize our understanding of the cosmos. If only scientists would listen. But after mailing and emailing and trying to talk with cosmologists, they have been met with silence.

We realized that in presenting the wonder and awe of science's picture of the universe, it is easy to lose sight of how we got here. The *act* of doing science was lost in the message. Science is a process by which an idea about the physical world is created, developed, and held up to our observations of nature, the ultimate arbiter.

In particular, we have often explained to aspiring revolutionaries this important principle: you have to explain *all* the data. You can't just pick your favourite and ignore the rest. And in modern cosmology there is a lot of data, and a lot of different types of data, to explain.

So, this book was born. We decided to shine a light on how scientists build and battle-test their ideas. We will focus on the critical observations that have laid to rest many cosmological contenders, as well as the challenges that are yet to be explained.

Keep up to date with further news, updates and commentary from the authors by following **@CRHandbook** and their YouTube channel, **AlasLewisAndBarnes**.

1 UNDERSTANDING SCIENCE

It is necessary to get behind someone,
before you can stab them in the back.

Sir Humphrey Appleby
Yes, Prime Minister (BBC), 1987

We want to teach you how to overthrow a scientific theory.

That might sound a little "anti-science", but actually you'll be doing scientists a favour. We learn something when bad ideas are exposed. Science often progresses by supporting the reigning ideas, but at other times it has been necessary to storm the castle and install a new monarch. That's how many great scientists rose to fame. *Vive la révolution!*

But you've got to do it right, and that's what this book is about. Revolutions fail for attacking the wrong target, following the wrong tactics, and underestimating the old order. Scientific theories are ideas about the natural world. They claim to know what the universe is like and how it behaves. This tells you how to dethrone a scientific idea: take up the weapon of *observations* and aim squarely at its *predictions*. Show that it can't handle the truth. And be ready with your new monarch when the throne is vacant.

To do all that, you must know your enemy. These wise words from Sun Tzu (or, if you prefer, Rage Against the Machine) are very relevant here: before you can launch a scientific revolution, you need to know the facts, and you need to know the ruling theory and its predictions. Theories aren't installed on the scientific throne by accident, so do your homework.

This book will hand you the facts, point you in the direction of the castle walls, and wish you the very best of luck. In particular, we'll be looking at the biggest scientific target of them all.

This book is about the universe.

It's about how we observe the universe, either with our naked eyes, or with the many telescopes that now survey the heavens, sensitive to radiation our eyes cannot perceive. But more than that, this book is about how we *understand* the workings of the universe, from its fundamental properties to its largest features. It's about how we put the pieces together.

Current scientific orthodoxy paints a picture of the cosmos that has been built up from many centuries of observation, experimentation, and hard thinking. Great minds throughout scientific history have laid the groundwork, carefully studying the basic rules of motion, space, time, atoms, light, and gravity, to provide the mathematical tools we need to comprehend the changing heavens. Today, *cosmology* – the study of the universe as a whole – is hailed as a paradigm of scientific success.

But what a strange picture! Many find modern cosmology completely unbelievable. The universe, we are told, was born almost 14 billion years ago in a hot and fiery event, cheekily named the *big bang*. At its beginning, everything was compressed into a point of infinite density and infinite temperature. In the aftermath, the universe is *expanding*, but it's not expanding *into* anything. Space itself is stretching.[1] Today, the galaxies we observe in the night sky all appear to be moving away from us. A vast sea of galaxies, stars, and planets fills this expanding space, but because light only moves so fast, most of this universe will be forever beyond the reach of our telescopes, over the *horizon*.

What about the stuff in the universe? Compiling an inventory would appear to be straightforward, if painstaking: just add up all of the stars, planets, and gas clouds that inhabit galaxies and the spaces between the galaxies. But cosmologists say that there is more to the universe than the stuff that we can see. Much, much more. A *dark side* of the universe, which we cannot touch or feel, dominates its energy budget and controls its expansion.

Firstly, modern cosmology tells us that there is *dark matter*. This stuff pervades every galaxy, holding stars in their orbits with its gravitational pull. But dark matter emits no light of its own, and so remains unseen by our telescopes. Stars illuminate the heavens, but dark matter accounts for more than 85% of the mass in the universe. The atoms that make up you and me, stars and planets, are little more than frosting on the cosmic cake.

And then modern cosmology tells us about *dark energy*, a substance as pervasive yet more elusive than dark matter. The case for dark energy was made only in the past few decades. We are told that this substance governs the dynamics of the universe on its largest scales, causing the expansion to accelerate, and driving us towards a cold, dark, dead future.

Why would anyone believe all of that?

A quick internet search turns up plenty of websites, blogs, and videos decrying modern cosmology as wrong, illogical, or even a conspiracy of the scientific establishment that suppresses voices of criticism. Modern cosmology, they claim, is a sham, purposefully distorted and hyped in the hunt for funding. Cosmologists are little more than a self-serving cabal, crushing all opposition.

Maybe, dear reader, you are one of these revolutionary voices, wanting to put science right. Maybe you have ideas about the laws of physics and how they impact our view of stars and galaxies. Maybe you have tried to engage with established astronomers and cosmologists to express your ideas and explain why their view is misguided, but have received a cold shoulder. Why are academics, locked up in their ivory towers, so sure they are right?

Our goal is to explain how physicists, astronomers, and cosmologists developed their picture of how the universe behaves, why they talk about it the way they do, and to tell you what you need to do to confront their strange ideas and begin a revolution. We'll help you build a strategy to battle modern science on a more even playing field, and to ensure that your voice is heard amongst the scientific din.

Just What Is Science?

Warning: the following discussion is very physics-o-centric!

To an outsider, science can be a difficult beast to understand. The media – and especially health advertisements – often tell us

"Science says ..." and "Scientists have discovered that ...", but science is not a single, monolithic enterprise. The scientific community consists of many thousands of individuals who often specialize in a narrow set of fields. Some scientists design experiments, some perform observations, and others wrestle with abstract mathematical theories. All spend far too much time in front of a computer. But what is the goal of science?

We begin with an important point: scientists try to predict the future.

If you are not familiar with the workings of science, this might seem a little strange. A flick through popular science magazines such as *New Scientist* or *Scientific American* will reveal stories that focus on *big* scientific questions such as "What is spacetime *really*?" and "What is quantum mechanics *really* telling us about the universe?" But we can't attack these deeper, foundational issues without some help.

In particular, it will help if we can bring these lofty questions down to a practical level. This is the part of science that plays "what if" games, constructing possible physical scenarios and teasing out implications. What if particles of light (*photons*) possessed a tiny amount of mass? What if a cloud of matter collapsed under its own gravity? What if I heat some hydrogen to 10 million degrees? Answering such questions requires more than a vivid imagination: we need our ideas to be translated into the language of mathematics. Sometimes, entirely new mathematical ideas need to be discovered and developed.

The goal of this precision is to connect our ideas to data. Can our new idea account for existing observations of the universe? And, just as importantly, are there any future observations that we could make that would provide further evidence for or against our idea? Can we get one step ahead of nature?

Take gravity as an example. In the 1680s, Isaac Newton published his incredibly successful theory of gravity. With one simple law, he explained how apples fall and how the planets move. Using Newton's law, Edmund Halley was able to predict the future motion of the comet that now bears his name. In 1705, he calculated that it would return in 1758. Sure enough, on Christmas day, it was spotted by a German farmer. Sadly, neither Newton nor Halley was alive to see it.

However, in the mid 1800s, Newton's theory was struggling. Astronomers had discovered that the innermost planet, Mercury, was orbiting slightly out of place, as if pulled by an unseen planet near the Sun. Some even claimed to have observed this newest member of the Solar System, which had been dubbed "Vulcan". Other astronomers, however, could not confirm this sighting. As evidence evaporated and Vulcan consistently failed to turn up where it was predicted to be, this mysterious shortcoming of Newton deepened into a crisis.

In the early 1900s, Einstein proposed his radical new theory of gravity – called the *general theory of relativity* – in which space and time themselves warp, stretch, and wobble. While Einstein's prediction of the orbit of Mercury is only slightly different from that of Newton, that was enough to beautifully align theory with observation. The planet Vulcan was banished to the scientific scrap heap.

Einstein's explanation of Mercury's orbit is impressive, but, like Newton's explanation of the motions of the planets, it comes after the data. We knew about the orbit of Mercury before Einstein proposed his theory. This is sometimes called a "post-diction".

Is there anything wrong with post-diction? We certainly can't discard all the evidence we found before a theory was proposed. Our scientific results would be swayed by something as contingent as what historical order we human beings happened to discover some idea or perform some experiment. That could depend on all sorts of irrelevant factors, like whether Thelma the Theorist took a few days off, or Xavier the Experimenter had a particularly good breakfast.

In principle, prediction and post-diction carry equal weight. But in practice we want to know whether a theory explains the data *naturally*, rather than being glued together from makeshift bits and pieces. Sometimes we can judge this by directly examining the assumptions that underlie the theory. But it is not always easy to tell. Predictions dispel this worry: you can't cook up a theory just to explain data if you don't have the data yet. If a theory correctly predicts the result of an experiment that we haven't done yet, then that is impressive.

So, when a new theory is proposed, we start asking "what if" questions. With Einstein's theory in hand, we have a whole new theoretical universe to explore. We look for new opportunities to test whether these ideas are correct. Einstein predicted that gravity would bend the path of light rays moving near massive objects. Famously, this effect was observed by the British astronomer Sir Arthur Eddington during a solar eclipse in 1919, confirming general relativity's predictions and propelling Einstein to further international fame.

Einstein's theory continues to make successful predictions. In 2015, a hundred years after Einstein's announcement of his new theory, scientists confirmed a hugely important prediction of general relativity: gravitational waves. Space and time can ripple. The discovery of these feeble vibrations, typically swamped by the everyday groans and grumbles of life on Earth, required half a century of effort to build an extraordinarily sensitive detector called the Laser Interferometer Gravitational-Wave Observatory (or LIGO for short). The results were spectacular, with the first signal revealing the merging of two black holes 3 billion years ago in the distant universe. LIGO has opened up a new window on the cosmos.

While Einstein's name is synonymous with scientific genius, you don't need to venture far into the outskirts of the internet to find many people who object to his ideas. Some play the man, rather than the ball, accusing him and the scientific community of outright fraud. Relativity is obviously crazy, they say, but it allows fat-cat scientists to keep feeding off the public purse. Others will decry the "logic" of relativity, often voicing a dislike of the notion of curved space and time, and even accusing the scientific establishment of wilful blindness to their unrecognized genius.

But science holds onto general relativity, not because of hero worship of Einstein, or because we are part of a secret conspiracy. Rather, we use his theory because it works. Physicists dream of proving Einstein wrong; we just haven't been able to do it. We are devising new ways to draw out predictions, and building new experiments to test those predictions.

As we said at the beginning of the chapter, the reigning monarchs of science didn't get there by accident. But they are always vulnerable, because every prediction is a chance to fail. So, what do you need if you want to revolutionize science? A new monarch. You need a model!

Just What Is a Model?

The word *model* has several meanings in the English language, and this can lead to some confusion when talking about a "scientific model". Anarchic comedian Alexei Sayle once said, "my girlfriend's a model. She's an Airfix kit of a Stuka dive bomber!"

We can understand the most important thing about a scientific model by thinking of a model house. Everything in the model is to scale, with one-twentieth size windows, doors, rooms, cupboards, and more. The useful thing about this model is that we can use it to answer questions about the real house. Suppose you want to know whether you can rearrange the living room to incorporate that new sofa you've had your eye on. You can answer this question with the model. If we make a one-twentieth scale model of the new sofa, then we can easily rearrange the model room to see if everything fits. For an accurate model, if the model sofa fits into the model house, then the real sofa would fit into the real house.

This is the crucial feature of a model: using the right translation, we can turn a problem in the real world (will the sofa fit in my living room?) into a problem in the model (will the model sofa fit in the model living room?). We then solve the problem in the model. If the model is an accurate representation of reality, then we have also solved the problem for the real world.

In the case of a model house, the translation between the model and reality is simple: it's just 20 times smaller. For a scientific model, the mathematical framework can be more complicated, but the crucial feature is the same: we can translate a question about the real world into a question about the model. Because we can relate between the two, we can make predictions. We can ask questions such as "what if I performed such-and-such experiment?"

Let's take another look at Newton's model for gravity. (We're physicists. We like Newton!) We can express his idea in words: gravity will produce a force between two masses, whose magnitude is proportional to the product of the two masses, and inversely proportional to the square of the distance between the masses. That's interesting, but not much use to a working scientist. To a scientist, the useful form of Newton's law of gravity looks like this:

$$\vec{F} = -G\frac{M_1 M_2}{r^2}\hat{r}$$

If you are not a fan of mathematics, and if this equation looks like little more than gobbledygook, don't worry too much. We can look at this like a machine, where we input two values for the masses, M_1 and M_2, and the distance between them, r, and this machine returns the gravitational force between them. The other number in the equation is G, which is known as Newton's gravitational constant. It scales the numbers so the result has the correct unit (which, for force, is the *newton*). Finally, $\hat{r}$ ("r" with a little hat) is known as a *unit vector*; it tells you that the force pulls the masses towards each other. But what can you do with this bit of mathematics?

We turn to Newton's laws of motion. We can state the idea in words as "forces cause objects to change their speed and direction of travel". But as we have noted, it's the mathematical version of the law that allows us to make precise predictions:

$$\vec{F} = m\vec{a}$$

This equation might be familiar from high-school physics; F is the force, a is the acceleration, and m is the mass. Combining these equations, we can start with information about the position and velocity (which encodes speed and direction) of the objects in the system at a particular time, and transform it into a prediction about the future of the system. For example, if we know where all the Solar System's planets are today, and how fast and in what direction they are moving, we can calculate where they will be at any future time.

The point of all physical models – Newton's, Einstein's, and anyone else's – is that we can ask questions about the universe. Given where I saw the planet Mercury last night, where will I see it tonight? By how much will the path of a light ray bend as it passes close to the Sun? We can ask Newton's model, and we can ask Einstein's model, and then we can actually look at the universe to see if either is correct.

The lesson is that if you are going to revolutionize science, you need a mathematical model. Words will not do. As scientists, we regularly get emails and letters espousing new ideas about the cosmos, from theories about fundamental particles to new interpretations of galaxy redshifts and the expansion of the universe. Surprisingly often, the author confesses that they are unable to express these ideas mathematically. I'm sure my idea is correct, they say, I just need some help working out the mathematics. To a scientist, and particularly to a physicist, this is a bit like saying "I have a great idea for a symphony; I just need some help with the musical notes" or "I'm sure I could do brain surgery; I just need some pointers on where to start cutting."

For a physicist, you don't really *have* a theory until you can think about it clearly enough to put it in mathematical form. Without precise predictions, it is too easy to fool yourself into thinking that the data is consistent with your idea. We need to predict measurements and observations, so that we can hold this mathematical model up to nature.[2]

What Makes a *Good* Scientific Model?

What does a scientist want in a scientific model? We have emphasized that your model must present a precise, quantitative picture of the universe, one that allows us to predict the results of experiments. But this is not the only criterion that scientists use. Historians and philosophers of science, by study-ing how scientists actually argue for and against theories, have proposed sets of *theoretical virtues*, that is, traits of a good scientific idea.

Not everyone agrees about all of the virtues, of course, but there is a common core that scientists will recognize. We will look at a recent list of twelve theoretical values (TVs) compiled by historian Mike Keas.[3] His list is helpfully comprehensive: while the twelve values overlap somewhat, each pinpoints something important about good scientific theories.

The first three relate to how your theory handles the evidence.

TV1. *Evidential accuracy:* your theory accounts for or fits the data well.

TV2. *Causal adequacy:* your theory posits causes that account for the effects we see in the data.

TV3. *Explanatory depth:* your theory applies to a wide range of scenarios.

Clearly, if your theory is correct, or at least approximately correct, then it should explain the data (TV1). All the data! Cherry-picking – focusing on the results that your mathematical model can describe, while ignoring those where it fails – is a scientific sin. This is a sure road to being ignored by the scientific community.

But scientists want more from a theory than this. The theory that the continents can move over the surface of the Earth explains why they appear to fit together like a jigsaw puzzle. But when it was first proposed, this theory was rightly criticized because it lacked causal adequacy (TV2): it didn't tell us *how* the continents moved. Frankly, no one had much of an idea of how something as large as a continent *could* slide around the Earth's surface. The theory of plate tectonics added the all-important details.

But the theory of plate tectonics does even more. It has impli-cations for a wide range of facts about the Earth's surface: how mountains form, how lava comes to the surface in volcanos, and the origin of earthquakes along fault lines. Scientists prefer broad theories that explain a lot about the universe (TV3).

The next three virtues are about how your theory hangs together.

TV4. *Internal consistency:* your theory does not contradict itself.

TV5. *Internal coherence:* your theory's various parts fit together neatly and naturally, with no internal tension or tacked-on assumptions.

TV6. *Universal coherence:* your theory sits well with other warranted principles.

Obviously, if one part of your theory contradicts another part, then it has self-destructed (TV4). But more generally, a theory can fail to hang together in a convincing way. It may need too many ad hoc bits and pieces, tacked on for no good reason. The paradigmatic example is *epicycles*: when the ancient Greeks observed that the planets don't move in perfectly circular orbits, Claudius Ptolemy proposed that the planets move along "circles on circles". There is no deep or natural reason for these epicycles. They explain the data, but in a clumsy way. Even Copernicus's model, which correctly placed the Sun at the centre of the Solar System, needed epicycles. A more coherent model of the Solar System would await the work of Johannes Kepler and Isaac Newton.

Also, if your theory violates established principles, then scientists are suspicious. Suppose your theory, despite fitting all the experimental facts, fails to conserve energy in certain circumstances. This is reason to worry since conservation of energy is a time-tested principle. We aren't going to discard, or even grant a few exceptions, to this principle on a whim.

The next three virtues are aesthetic.

TV7. *Beauty:* your theory strikes scientists as beautiful.

TV8. *Simplicity:* your theory explains the same facts with fewer starting assumptions.

TV9. *Unification:* your theory explains more *kinds* of facts than rivals, relative to its starting assumptions.

TV7 might surprise you, but a long line of important physicists can be marshalled in support. The physicist Paul Dirac went as far as saying, "it is more important to have beauty in one's equations than to have them fit experiment." Henri Poincaré

spoke of "the intimate beauty which comes from the harmoni-
ous order of its parts and which a pure intelligence can grasp".
In a letter to Albert Einstein, Werner Heisenberg stated the
following:

> That these interrelationships display, in all their mathematical
> abstraction, an incredible degree of simplicity, is a gift we can only
> accept humbly. Not even Plato could have believed them to be so
> beautiful. . . .
>
> You must have felt this too: the almost frightening simplicity and
> wholeness of the relationships which nature suddenly spreads out
> before us and for which none of us was in the least prepared.[4]

Why would scientists – hard-nosed, no-nonsense, just-the-facts
people, supposedly – be concerned with something as subjective
and nebulous as beauty?

One important reason is found in physics as we know it. Many
of the theories of modern physics, when one can fluently speak
the appropriate mathematical language, are strikingly elegant,
symmetric, and ingenious. Einstein's general theory of relativity
is a great example. The central insight – gravity is the warped
geometry of space and time – is a piece of creative brilliance,
arguably the greatest single theoretical insight in the history of
physics. The connection of this idea to the beautiful mathematics of curved spaces seems almost inevitable; in the appropriate
language, the theory is stated very simply, even poetically.
The range of observed data that the theory explains is enormous.
Within one concise equation, sitting innocently on the black-
board, one feels the weight of a million worlds to be explored –
expanding universes, spinning black holes, slowing clocks, bending light rays, gravitational waves, and even the possibility of
time travel.

But what exactly does a physicist mean when they say that
a theory is beautiful? Most physicists, we think, would say that
they don't exactly know.[5] As with all experiences of beauty, it is
vivid and immediate, but not easy to describe. The other
theoretical virtues try to explain a bit more.

One common feature of mathematical beauty is simplicity. The scientific enterprise generates an extraordinary amount of data every day. A single night at a large telescope will easily fill a computer hard drive, and that's just astronomy. One of the lovely things about beautiful theories is how succinctly they can be stated: a few postulated kinds of entities and laws are all you need to explain a mountain of data. It's like cracking a code or solving a puzzle.

Another important feature is *unity*. Beautiful scientific theories give a sense of how the whole of nature fits together into one grand picture. Ptolemy, in the second century AD, glimpsed something wonderful in his theory of the Solar System:[6]

> *I know that I am mortal by nature and ephemeral, but when I trace at my pleasure the windings to and fro of the heavenly bodies, I no longer touch earth with my feet. I stand in the presence of Zeus himself and take my fill of ambrosia.*

When a theory unifies our view of nature, it explains many seemingly disconnected facts. It explains how nature holds together. (Part of the reason for the beauty of our scientific laws, of course, is that they describe our beautiful universe.)

The final set of theoretical virtues concern how a theory fares over time, as it is examined, extended, and utilized by the scientific community.

TV10. *Durability:* your theory has survived testing by new experiments and new data.

TV11. *Fruitfulness:* your theory has pointed to new discoveries, such as successful novel prediction and unification.

TV12. *Applicability:* your theory has led to the development of new technology.

If your theory has really peeked into the inner workings of nature, then it should continue to illuminate the way the universe works. It should be able to explain new data that we gather (TV10).

As we discussed above, the advantage of prediction – theory first, observation second – over post-diction is that we can be sure that the theory wasn't jerry-rigged to explain the data. It is particularly impressive if (TV11) a new theory suggests an experiment or observation we hadn't even thought to make. Physicists love this: it gives us something to do.

One particular form of evidence for a new theory of nature is whether it can be put to technological use (TV12). This isn't true of all theories, of course: working out how a galaxy makes its stars isn't going to lead to star-making factories any time soon. But if we really understand how electrons and gravity work, then this should help us design and build devices that make use of this knowledge. We understand electrons well enough to build CRT televisions, and we understand gravity well enough to put satellites in orbit.

We note again that while these criteria are overlapping, each captures something important about what we expect from a good scientific theory. Our best theories fulfil most of them. But if your theory is all numerical coincidences and no physical insight, if it lacks unifying principles, if it ignores huge swathes of data, if it is made from an odd assortment of unconnected assumptions, if it is mathematically mundane, if it doesn't seem to produce new predictions or new ways of looking at the physical world, then scientists will have a sneaking suspicion that there's something missing.

The Real Process of Science Part 1: Publishing

So, now you have your theory. You've thought through its foundations, its implications, and its applications. You have a clear vision of the real nature of the universe. You're sure it ticks off enough of the theoretical virtues to be taken seriously. You're ready to explain your discovery to anyone who will listen. How do you make scientists pay attention?

Some advice: writing letters or emails to prominent scientists probably won't work. They receive too many of those. If well-known scientists responded to all the supposedly revolutionary ideas that landed in their inbox, they'd have time for little else.

A scientist – even a famous one – might answer a concise question if you're polite, especially if you open with, "I read your book/paper/article and I have a query." But your glorious 400-page unsolicited Word document, with seven different fonts, Microsoft Paint diagrams, and ALL CAPS, is headed straight for the trash. We haven't studied this phenomenon systematically but, in our experience, there is a remarkably strong correlation between how utterly bonkers an idea is and how many fonts are used by **its** *defender*. Even too much **bold** and *italics* is a red flag. Don't do this. OR UNLEASH YOUR INNER CAPS LOCK. Or use multiple colours. Or write a webpage that is one enormous paragraph. Good typesetting is subtle. If you really want your document to look scientific, learn to use a program called LaTeX.

Raging in internet chat rooms and blog comments won't achieve much either. In fact, scientists are unlikely to hear of your idea even if you write a book about it; new theories in physics are rarely first published in a book.

Popular culture often talks about the scientific method as an idealistic "wash, rinse, and repeat" procedure: you have an idea, test the idea in an experiment, then accept or reject your idea and start again with a new idea or experiment. Real science isn't quite this neat. Different fields approach nature differently. But there is at least one key common element to the practice of science: publication.

Publications in a scientific journal are the currency of science. On the résumé of a scientist, there will be a list of their publications and the journals in which they appear. Why are journal publications so highly regarded? Because this is the *first step* of peer review by the broader research community. To get published in an established journal, a manuscript is submitted to an editor. The editor takes the first glance at a paper, just to check that it looks like, well, a paper: is it laid out appropriately, does it present a background, approach, methods, results, and discussion, and does it look like it makes at least a little bit of sense? For articles that overcome this low bar, the editor will then seek comment from external referees.

If it passes their review, sometimes after a few revisions, then it is published.

Scientists have a choice about which journal they send their paper to. We look for a good reputation, proper editorial oversight, and peer review by other scientists. The hierarchy of journals is sometimes ranked by something called *impact factor*, which measures how many times (on average) a paper in that journal is cited over a particular time period. It's an awful statistic, strongly skewed by individual papers that are very influential and heavily cited, shining glory onto other articles that share the same pages. This discussion of journals, impact factors, and gaming will cause some of our colleagues' blood to boil. Scientists don't particularly appreciate reducing their work down to a blunt statistic. Still, impact factor is easy to calculate.

High-impact journals such as *Nature* and *Science* cover all of science and reject most papers that they receive. Well, you can't be exclusive without excluding. They require submitted work to be highly novel and innovative. This doesn't mean that it is right; in some ways, it means that the work is riskier.

Then there are the "bread-and-butter" journals in a scientific field. In astronomy, for example, most papers are submitted to journals such as the *Astronomical Journal*, *Astrophysical Journal*, *Monthly Notices of the Royal Astronomical Society*, and *Astronomy & Astrophysics*. These journals accept papers from all over the world. While their refereeing is robust enough to see many papers rejected, they aren't necessarily looking for the latest sensation. As well as these larger journals, smaller journals can be found attached to national societies or observatories. These provide a place for less impactful research, conference proceedings, technical reports, or student projects with an interesting result.

The reward for publishing in a higher impact journal is a more impressive CV for a scientist and a boost for their university up the rankings. As a result, it is not unheard of for a paper to be submitted initially to the highest impact journal, only to be rejected and be resubmitted to a lower ranking journal, and so on.

The lesson is that merely getting your paper published isn't as important as *where* it is published. Scientists will take note of where your paper has appeared, as this will signal the level of refereeing and editorial review it has received. If you claim to have revolutionized cosmology, waving your published paper as evidence, it had better be published in a high-ranking physics or astronomy journal. Appearing in the *Bulgarian Journal of Basket Weaving* will not garner much attention.

The Real Process of Science Part 2: Peer Review

As part of its assessment at a journal, a submitted paper will be sent out to *referees*: experts in the field who provide detailed comments on the paper, in particular identifying any glaring mistakes. Referees also identify whether the work is interesting, significant, and potentially of use to the scientific community. Papers must pass this interrogation to be published.

But don't overestimate the importance of a thumbs-up from a referee. A passing grade does not mean that the paper is correct, or has been accepted as immutable scientific orthodoxy. What it means is that some scientists have judged that it is not obviously wrong and probably of use to other researchers.

And don't think that the assessment of an article for a journal by referees is all there is to peer review. It is merely *the first step* in assessment by the scientific community. Peer review continues long after an article has been published, in the continual assessment of whether the idea is interesting, accounts for new data, and spurs new ideas. The most definitive indicator of scientific impact, craved by scientists and by university bean-counters equally, is the *citation*.

When scientists list their journal papers, they usually also list the number of times their article has been cited in someone else's work. Why? Because this means that others in the scientific community have studied their work and found it useful. Oscar Wilde's quip that "there is only one thing in life worse than being talked about, and that is not being talked about" is true for scientific papers. The worst thing that can happen to a scientist is that nobody takes notice of your work. At best, no-one has realized its importance *yet*. This can happen: in 1967,

Steven Weinberg published a theory in particle physics for which he would later win the Nobel Prize.[7] But in the first four years after publication, it was only cited twice. Once his colleagues realized its importance, however, this soon changed. It has been cited, on average, four times a week for the last 50 years, for a total of 11,000 citations.

But let's be honest. A lack of citations is more likely to be a sign that it was not an interesting or good piece of work.

Citations are an important measure of scientific impact. Universities and funding bodies are continually assessing the research activities and output of their staff. They are glued to international university rankings like a football fanatic glued to the league table; positioning in the table brings renown and prestige, and with it, better chances at securing research funding and foreign students. Universities and funding bodies want impact to be immediate.

Scientists, however, tend to have mixed feelings about citations. Those of us engaged in so-called *blue-sky* fields, which study fundamental questions about the structure of matter and the universe at large, know that it can take a long time for some research to be appreciated and result in scientific impact. Like Weinberg's paper, plenty of good scientific work in astronomy and cosmology, such as the prediction of gravitational lensing or gravitational waves, or the initial observational clues of dark matter, waited for many decades before its importance was realized.

At this point, we tip our hats to our mathematician cousins. Physics is grateful for the constant flow of new mathematical ideas that allow us to understand the workings of the universe. However, advances in mathematics may lie around for centuries before being noticed for what they are. This means that mathematicians can die unrecognized in their lifetime.

The Real Process of Science Part 3: Presentation

As well as publishing material in recognized journals, scientists spread the word about their work by presenting at conferences and workshops. These are quite varied occasions, some with a handful of people, some with hundreds, some covering broad areas of astronomy and cosmology, others focused on very specific topics.

For example, these words are being written on a Boeing-737, winging its way to Cairns in northern Queensland, Australia. The authors are heading for "Diving in the Dark", a conference about the dark side of the universe, both dark matter and dark energy. While at the airport, we discovered that Cairns is hosting a second astronomical meeting this week, specifically focused upon the centre of the Milky Way galaxy. At each meeting, astronomers will present, discuss, and argue about the latest research, share new ideas, and forge new collaborations. Walking away from such meetings, an astronomer will have a feel for the latest research in their area, have gained new ideas of where to take their research and, importantly, have promoted their own ideas to the community.

No matter how lucid your writing, you just can't beat a good audio-visual presentation for conveying the big picture of your idea. Talking about your latest research at major meetings makes it much more likely that other researchers will read your papers. The goal is not just appreciation: you want them to run with your idea, thinking of new implications and new ways of testing those implications. When they write their next paper, they may even give you one of those treasured citations.

As with publications, presentations should be at significant conferences that highlight the latest data and results. Wowing the monthly meeting of your local Quidditch society doesn't do much for your theory's credentials. When you're a young researcher, you may only get 5 or 10 minutes, so have an elevator-pitch version of your ideas ready!

Does this sound like a lot of hard work? It is. And that's the way it should be. Revolutions shouldn't be easy!

Most importantly, be prepared to fail. In fact, accepting that you are wrong and moving on is a key attribute of a good scientist. Sticking to your guns when all the data says that your idea is wrong is the stuff of pseudoscience and irrelevance.

Admittedly, the evidence against a theory isn't always completely definitive, so keeping an option on the table can be
appropriate. But, as evidence mounts and the scientific community moves on, ideas that can't keep up are continually being
discarded.

Does Science Want a Revolution?

Isn't this all a bit optimistic? It assumes that scientists are, at
least on average, rational, reasonable, and unbiased, immune to
cognitive distortions such as groupthink, confirmation bias, and
lack-of-coffee. But scientists are human beings, too. Maybe
they're just doing whatever it takes to get another research
grant or win the approval of their peers.

In particular, if scientists *wanted to* suppress new ideas, then
the mechanisms we have described – journals, editors, peer
review, and citations – seem to provide an all-too-easy way to
do it: you tread down an idea by preventing it from being
published. Have reviewers reject it or, even more effectively,
have the editors of the journal reject it before it is sent out for
review. Without the minimal stamp-of-approval from a journal,
most of the time-starved scientific community won't consider it.
And so, an idea – even a good one – can be suppressed by the
scientific illuminati.

Now, we can't discuss every supposed case of scientific cen-
sorship, nor do we want to defend everything that every scientist
or journal editor has ever done. We cannot deny that the history
of science is littered with theories that took too long to die,
cherished by scientists who fought hard to ignore the evidence
that the time for their pet theory had passed. This is not a new
phenomenon, as a great scientist of the early twentieth century,
Max Planck, was reputed to say, science progresses "one funeral
at a time".

We're formulating a game plan here, not settling every griev-
ance. What's important is that while science, like every human
activity, isn't perfect, we can make these imperfections work
for us.

Firstly, as a scientific revolutionary, you must realize that cranks exist, as do nuts, loons, and buffoons. In addition, there are well-meaning amateurs without the breadth of knowledge or mathematical ability to contribute meaningfully to the scientific enterprise. Explaining to each of them why they are wrong would be a colossal waste of time. More scientific and would-be scientific work is published every day than any person can read and digest. If you want to be noticed, you'll have to do *something* to distinguish yourself from people that scientists have good reason to ignore.

Secondly, most scientists acknowledge an obligation to communicate their results to the public. For example, scientists who study the natural world will want to inform people (and their elected representatives) of threats to the environment or the survival of a particular species. Many scientific advances have potential technological applications, which will need to be explained to potential investors and customers. Astronomy, in particular, is driven by the human thirst for knowledge, rather than being directly aimed at new technology. We are mostly supported by government research grants, so having the tax-paying public share our curiosity about the universe is in our best interests.

If scientists expect their ideas to be respected by the layperson, we cannot act like a secret society. We must show openness to new ideas, or, at the very least, not tamper with the scientific ecosystem: reason and evidence, peer review and citations. If their taxes pay our salary, we *owe* (to some extent) the public an explanation of why we prefer our ideas to theirs.

Thirdly, what if my new ideas threaten the old order and their lucrative research grants? Scientists, so the story goes, are up in their ivory towers, with their cash and their students, wanting little to change. Surely great ideas are continually suppressed by the "Establishment" to stop them from upsetting this cushy apple cart?

But this story doesn't hold together: you don't get new grant money for doing old science. When a scientist applies to a funding agency for a grant to support their research, they must propose to do something *new*. Similarly, PhD students need a novel project to work on, something that they can make their own, that no one has done before. Scientists are constantly on the lookout for a *hook*, a new idea or method, which they can present in a grant application and say, "look at this new thing that I can do!" So, show us how your new scientific idea could *contribute* to our case for more funding.

Finally, if you ask a young scientist why they chose this career path, or ask an older scientist what the most satisfying moment of their career was, they will often point to the excitement of a new discovery or insight. In the words of physicist Ed Hinds, "those of us engaged in scientific research generally do it because we can't help it – because Nature is the biggest and most complicated jumbo holiday crossword puzzle you have ever seen." No one becomes a scientist so that they can plod along behind the establishment. Scientific revolutions get the Nobel Prize.

These forces – earning public respect, winning funding, and advancing knowledge – keep science from stagnating and keep us open to new ideas. This is an important lesson for the cosmic revolutionary: scientists don't want just any old revolution. They want one that offers new ideas and new directions, that creates opportunities and deepens understanding.

So *that's* your hook. What's your great insight, and how will it help us understand this marvellous universe? Give us something that can be held up to nature, firmly founded in mathematics. Give us a deep insight into an old puzzle. Uncover an unseen simplicity in our data. Point the way to new experiments that will test your idea. If you can do this, revolution is at hand.

How to Read This Book

This book examines the universe in discrete pieces, considering key observations that tell us something deep about the workings of the cosmos. We will endeavour to present these observations in as *raw* a form as possible, undigested and untouched by theory. This isn't completely achievable, unfortunately. But in astronomy we can at least focus on the question: what have our telescopes recorded?

We can report, for example, that the planets appear to move with respect to the fixed stars, and that some stars get brighter and dimmer at regular intervals. When we use a telescope that can measure the temperature of a light source, we pick up a strong signal all across the sky at almost three degrees above absolute zero. The interpretation of these facts, that is, the story of the universe that ties them all together, is up to you and your revolutionary idea.

Our aim is to provide the facts that need to be explained. We will give the most up-to-date versions of the key observations. This is important: we have noticed that many would-be revolutionaries focus too much on the *first* evidence that was presented for a certain claim. For example, they spend their energy critiquing Edwin Hubble's observations in the 1920s of a sample of 30 galaxies, completely ignoring the millions of galaxies we have observed since then.

Throughout, we will give references in the endnotes to the scientific literature regarding both observations and theoretical interpretation. While papers in scientific journals are often hidden behind a paywall, many can be found for free at *arxiv. org* with a quick search. Most scientists, if you ask them succinctly and politely, would be happy to email you a copy of one of their papers.

Remember: if your scientific ideas can account for all of the observations presented in this book and provide future predictions that can be held up to nature, then scientific legitimacy awaits. Alas, fame, fortune, and a lucrative book deal are not guaranteed.

Good luck.

2 HOW DARK IS THE NIGHT?

At night, the sky is dark. If this is new information for you, the rest of this book is going to blow your mind.

Some very surprising conclusions follow from this familiar fact. The vast silence of the night sky has led poets to dream of infinite space, timeless and still, filled everywhere with ageless stars. However, our universe cannot be like that. To see why, we need to trace the ideas and the evidence that lead to Isaac Newton's universe.

The World before the Revolution

The original cosmic revolutionary was Nicholas Copernicus (1473–1543). The publication of his *De Revolutionibus Orbium Coelestium* (On the Revolutions of the Heavenly Spheres) in 1543 marks the end of the dark ages, the banishment of old ideas about humans being at the centre of the universe, and the beginning of the scientific era. While Copernicus's work was resisted and suppressed by the religious forces of the time, eventually sense and, more importantly, evidence won the day.

Actually, no. That's the modern myth. For the last hundred years, says historian of science Allan Chapman, scholars have been systematically unpicking the idea of a dark age during which science was systematically suppressed.[1] The Middle Ages saw the establishment and advancement of the university, where every future physician, lawyer, and clergyman in Europe was required to learn logic, geometry, algebra, and astronomy, informed by a synthesis of the recently rediscovered Greek science and more modern thought. Isaac Newton (1642–1726), the greatest physicist of the scientific revolution (and arguably of all time), acknowledged his debt to his predecessors:

If I have seen further, it is by standing on the shoulders of giants.[2]

The eminent historian of science Edward Grant concludes his 2007 book *A History of Natural Philosophy* with these words:[3]

> *Modern scientists are heirs to the remarkable achievements of their medieval predecessors. The idea, and the habit, of applying reason to resolve the innumerable questions about our world, and of always raising new questions, did not come to modern science from out of the void. Nor did it originate with the great scientific minds of the sixteenth and seventeenth centuries, from the likes of Copernicus, Galileo, Kepler, Descartes, and Newton. It came out of the Middle Ages from many faceless scholastic logicians, natural philosophers, and theologians. . . . Without the level that medieval natural philosophy attained, with its overwhelming emphasis on reason and analysis, and without the important questions that were first raised in the Middle Ages about other worlds, space, motion, the infinite, and without the kinds of answers they gave, we might, today, still be waiting for Galileo and Newton.*

Science did not start with Copernicus. (As a tip to any aspiring revolutionaries: the tactic of proclaiming that everyone who came before you was a complete dunce has had, at best, mixed success over the centuries.) But he did revolutionize the way we look at the universe. So, what did the universe look like before the revolution?

Aristotle's Physics

As is so often the case, we begin with the Greeks. The philosopher Parmenides (515–450 BC) posed one of the most important philosophical problems of his age, which concerned the idea of *change*. Think about a caterpillar that changes into a butterfly. Is it still the same insect? You might say yes: it's the same matter, albeit rather spectacularly rearranged. Or you might say no: yesterday there was a caterpillar, and today there is a butterfly. Two things with different colours and structure can't be the same thing. Parmenides argued that this confusion exposed a logical problem with the very idea of change. If something changes, it "ceases to be itself", but this is absurd. The same thing can't become a different thing. So, change must be an illusion.

You might be thinking, "That's ridiculous. Everyone knows that things can change." But that's not the point of the exercise. It is easy to convince yourself that something must be wrong with Parmenides's argument; the challenge is to explain *why* it is wrong. Perhaps, when we've looked into the available accounts of change, we'll conclude that they aren't sufficient, and that Parmenides was right: change is an illusion. Ironically, Parmenides will have changed our minds. Or perhaps we'll agree with Heraclitus, fellow Greek philosopher, who argued that there is *only* change: "You cannot step in the same river twice."

Aristotle (384–322 BC) proposed a third way by making a common-sense distinction. As well as the way that a thing *is*, there are the ways that a thing *could be*. Change isn't nothing becoming something, but instead a *potential* something becoming an *actual* something. The caterpillar has no wings, but it has the *potential* to grow wings. When it undergoes metamorphosis, it becomes an insect that *actually* has wings. Similarly, an ice cube is not actually a pool of water, but it is *potentially* a pool of water. The ice cube does not cease to be itself when it changes; the application of heat actualizes its potential wateriness.

These potencies are not just in our imagination, because then anything could turn into anything. A caterpillar and an ice cube have potencies that are grounded in their *nature*. I can imagine ice melting into a puddle of water, and I can also imagine ice spontaneously growing wings. But, alas, the winged ice cube never happens. Similarly, the caterpillar never melts into a puddle of water.

The Aristotelian concludes that it is in the *nature* of a caterpillar to become a winged butterfly, and it is in the *nature* of ice to melt into water. Just as importantly, it is not in the nature of ice to grow wings, and it is not in the nature of the caterpillar to melt into water. For Aristotle, this explains why the cosmos is orderly rather than chaotic – natural things don't just behave as they please.

These potencies are not enough to produce change – they must be actualized. In particular, they must be actualized by something that is already actual. Those potential wings can't drag themselves into existence; that requires the action of an *actual* caterpillar. If ice could melt itself (regardless of its environment) then there would never be unmelted ice in the first place.

For the Aristotelian, then, the project of understanding the natural world consists of several key questions. What is the *nature* of the substances we find in the natural world? What actually exists, what are its (real, not imagined) potencies, and how are they actualized in the real world?

Aristotle Looks to the Heavens

With that framework, Aristotle looks out at the universe. He is soon presented with a puzzling contrast. On the one hand, things on Earth tend not to move very much, or very easily. If you look around the room you are in, you'll see that most of the objects are sitting still. If you kick something, it will move, but not for long. Cars need petrol, sprinting after the bus is exhausting, and that fridge isn't going to move itself.

Meanwhile, above our heads, the silent stars go by. Day and night, the Sun, Moon, planets, and stars just keep going around. They trace out – as far as Aristotle knew – perfect circles in perfectly smooth, uniform motion. What explains this motion?

The ancients had considered the idea that the motion of the heavens is only apparent, that it is the result of the Earth's rotation. But they ruled out this hypothesis with an experiment. Thanks to Eratosthenes (276–195 BC), they had a reasonable idea of how large the Earth was – and, yes, *they all knew that the Earth was round*. Then, if the Earth rotates once a day, we are all now and always have been rushing eastward at over a thousand kilometres per hour.

Here's the experiment: drop a stone down a well. The sides of the well are moving eastward at enormous speed, so surely the stone will hit the sides before it hits the bottom. But it doesn't. It hits the bottom, as if the Earth wasn't moving at all. This proved to the ancients that the Earth was stationary. (Only after Galileo, who explained relative motion and inertia, was this argument able to be countered.)

In light of observations of nature, and in the framework of potencies and actualities, here is how Aristotle put it all together. The Earth is made of four elements: earth, water, air, and fire. Each has a natural place in the universe and tends to move to that place. It is its *nature* to move towards that place. (This is a real tendency in the present, not an imagined future state.) The element earth tends to the bottom, and then water tends to sit on top of earth, air above the water, and fire rises to the top. So, for example, bubbles naturally rise in water, and earth naturally sinks in water. When the natural order is upset by *violent* motion – like throwing a rock in the air – the inherent natures of the elements tend to restore order. So, the rock will fall towards to the ground. Upon reaching the ground, it's in its natural place and will come to rest.

The Earth is a place of violent motion that is being damped down, a place of change and unrealized potential where things aren't quite in their proper place. In this picture, we are *not* at the exalted centre of the universe. We're at the bottom, where all the imperfections collect and coagulate. In Aristotle's universe, we are – in the words of Galileo in 1610 – "the sump where the universe's filth and ephemera collect".[4]

The heavens, however, are above all this imperfection. The heavenly bodies are changeless, executing their perfect circular motions with no sign of slowing or deviating. They don't have a tendency to move towards a certain place, like the four elements do. They just keep going around. So, they must be made of something other than the four elements. And, moreover, they must maintain this motion by contact with a perfect source of motion.

So, the planets and Sun move on crystalline spheres made of aether, the fifth element. And to make the entire universe rotate, there is an outermost sphere that moves all the others but is not itself moved. Aristotle called it the *prime mover*.

This was Aristotle's world, and his ideas reigned for two millennia.

Aristotle's writings cover much more than physics. He wrote hugely influential works on logic, rhetoric, poetry, mathematics, ethics, and politics. In addition, his approach to studying the world – actuality and potency, causes and essences – was extended to geology and biology. He was a skilled and voracious observer of the animal kingdom.[5]

Why have we taken the time to explain Aristotle's physics? Because, to Copernicus in the sixteenth century, Aristotle was the establishment. And you don't get to the top without explaining a few things. For two thousand years, Aristotle's system worked. It explained a lot about the natural world – why rocks fall, why fire rises, why the stars go around, and why violent motion doesn't last.[6] It did this in the context of a comprehensive worldview that extended from the foundations of knowledge to physical science to art, politics, and poetry.

But it was not perfect. As early as the sixth century AD, John Philoponus noticed that – *sometimes* – heavier objects don't fall faster than lighter ones, as Aristotle's system predicts. We emphasize *sometimes* because, as a matter of simple observation, heavier objects often fall faster. Try it: drop a leaf and a stick. The heavy stick falls faster. As modern physicists, we invite our students to imagine the experiment in a vacuum. In the absence of air resistance, leaf and stick fall at the same rate. But Aristotle was interested in natural motion, and the instruction to remove all the air seems like an unnatural thing to do. In the real world, bodies fall in air, and physics is supposed to explain the real world. (In any case, Aristotle thought that a vacuum was a physical impossibility.)

However, Philoponus's insight is perfectly valid; as thinkers in the Middle Ages rediscovered it, they tried to tinker with

Aristotle's physics. However, with the benefit of hindsight, we know that a revolution was needed.

Copernicus, Revolutionary

Born in Poland in 1473, Copernicus's remarkable career included studying canon law and medicine, working in politics and the military, and publishing essays on economics and classics. But his real passion was astronomy. He taught himself Greek astronomy from the influential textbook *De Sphaera Mundi* (On the Sphere of the World), written by Johannes de Sacrobosco in 1220. He studied Greek mathematics and pored over tables of astronomical data. While at university in Bologna, Italy, he befriended a professor of astronomy and began making his own observations.

Sometime around 1514, he began telling people about his revolutionary idea. About 30 years later, on his deathbed, he was able to hold in his hands the finished product of his work: *De Revolutionibus Orbium Coelestium* (On the Revolutions of the Heavenly Spheres).

His book was widely read – we have several hundred surviving early editions, many with extensive notes in the margins from their owners. Why were so many people interested in an astronomical theory? At least part of the reason is that many people had been trained in the basics of astronomy.

Throughout the Middle Ages, the standard university curriculum included astronomy. In particular, every clergyman – from the village vicar to the archbishop – needed to be sufficiently familiar with the heavens to calculate the date of Easter. The educated classes of Europe were familiar with and even dependent on Greek astronomy, and so could watch Copernicus's revolution with a great and well-educated interest.

Copernicus's system was advertised primarily as a calculating device. Since Ptolemy (AD 100–170), Aristotle's system of spheres had become complicated. It was discovered that the planets didn't travel in perfect circles but could be more accurately described using epicycles – circles travelling on circles, like a wheel on a wheel. This made calculation of planetary orbits difficult. A great many people welcomed a simpler way to keep track of the calendar.

Copernicus's revolution was to place the Sun at the centre of the Solar System, with the planets in orbit around it. However, Copernicus still had planets travelling on circular paths. This isn't quite right, as we now know, and so Copernicus's system didn't escape the need for epicycles. In fact, his system required about as many epicycles as Ptolemy's system.

But Copernicus's system has important advantages over Ptolemy's. It more easily explains a few odd features of planetary motion. If you watch Mars in the night sky, it will usually wander in a particular direction relative to the stars. (The word planet means "wanderer" in ancient Greek.) However, every now and then, Mars will wander backwards. An example is shown in Figure 2.1: pictures were taken of the same part of the night sky at 5 to 7-day intervals between October 2011 and July 2012. Ordinarily, Mars – the bright spot in the photos – would just wander from right to left, west to east. (Remember: east and west are reversed when you look upwards.) But, as you can see, for a few months, Mars goes from left to right, east to west! This is called *retrograde* motion, and Jupiter, Saturn, Uranus, and Neptune do it as well.

Now, Ptolemy's system can explain this, but only by jerry-rigging the epicycles. If we arrange just the right cycles at just the right times, Mars will briefly move backwards, as observed. And if we line up the other epicycles just right, we can make Jupiter, Saturn, Uranus, and Neptune move in retrograde as well. Just as there is no real reason for epicycles in Ptolemy's system, there is no real reason for them to conspire in this way.

In Copernicus's system, retrograde motion is inevitable. It happens whenever the Earth passes Mars in its orbit (Figure 2.2). Like overtaking a car in the inside lane, Mars will briefly appear to move backwards. Inevitably, Jupiter, Saturn, Uranus, and Neptune – the outer planets – occasionally appear to move backwards as well.

Take another look at Figure 2.2: when retrograde motion happens, Mars, Earth, and the Sun align. Think about standing

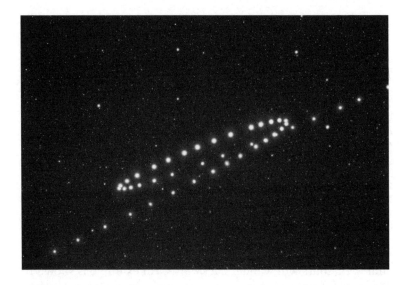

Figure 2.1. A collection of photographs that trace the path of Mars on the night sky from late October 2011 (top right) to early July 2012 (bottom left). The images are about 6 days apart. Mars shows retrograde motion, appearing to move backwards with respect to the stars. (Used with permission, image Credit & Copyright: Cenk E. Tezel and Tunç Tezel, The World at Night: www.twanight.org.)

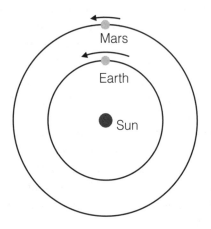

Figure 2.2. Earth overtaking Mars in its orbit, which explains the retrograde motion of the Red Planet (see Figure 2.1).

on the surface of Earth, at the point closest to Mars: it's midnight, and Mars is directly overhead. This is when we observe retrograde motion to happen, for all the planets. It's inevitable for Copernicus, but another unexplained coincidence for Ptolemy.

There is a second fact about the night sky that Copernicus naturally explains. Mercury is always seen within 28 degrees of the Sun, and Venus is always seen within 49 degrees of the Sun. There is no corresponding fact for the other planets – depending on when you're reading this, Mars could be anywhere relative to the Sun. Why?

For Ptolemy, more coincidences. But for Copernicus the answer is again very natural. Mercury and Venus are *seen* close to the Sun because they *are* close to the Sun. They are the planets *inside* the Earth's orbit. The other planets, being outside the Earth's orbit, have no such restriction.

Building on Copernicus's ideas, Johannes Kepler (1571–1630) was able to expel epicycles from the heliocentric model of the Solar System. He proposed a picture of the Solar System that later scientists would boil down to three simple laws of planetary motion.[7]

Kepler's first law concerns the *shapes* of the orbits. Kepler discarded circles and proposed that planets move along a slightly flattened circle called an *ellipse*. Figure 2.3 shows two examples: a slightly elongated ellipse on which a planet travels, and a more elongated ellipse for a comet. Crucially for explaining observations of the Solar System, the Sun is not at the centre of the ellipse. It is shifted to one side; the more elongated the orbit, the more the Sun is shifted.

Kepler's second law describes how the planets speed up as they get closer to the Sun. Look at the comet in Figure 2.3: Kepler tells us that the comet will be travelling faster on the right side of the orbit as it sweeps past the Sun, and slower on the left part of the orbit. Specifically, when the comet moves twice as far from the Sun, it is moving half as quickly.

Kepler's third law describes how fast the planets move relative to each other. A planet that is (on average) four times further

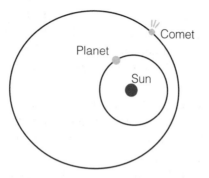

Figure 2.3. Examples of orbits around the Sun in the Solar System. Planets travel on almost circular orbits, while comets travel on elongated ellipses. The Sun is not at the centre; it's at a special point called the *focus* of the ellipse.

away from the Sun will take eight times longer to complete its orbit. Mathematically, the rule is that the square of the period of the orbit is proportional to the distance cubed. So, the outer planets not only have further to travel to complete their "year" around the Sun, they are also travelling more slowly.

Kepler's system has several advantages over Ptolemy's. Instead of over 50 epicycles cobbled together without rhyme or reason, we have three simple rules applied to each planet. And it keeps the advantages of Copernicus's insight: retrograde motions are explained, as is the closeness of Mercury and Venus to the Sun.

However, while Kepler's system is mathematically simpler, its physics is abominable ... if you're an Aristotelian. What's pushing the Earth around in its orbit? Where is its crystalline sphere? How can earth – dirt! – be attached to a crystalline sphere? The Sun is made of aether; how can it be at the bottom of the universe? If Kepler is right, then the Earth's equator is spinning at 1,000 kilometres per hour. And the Earth is orbiting the Sun at 30 kilometres per second! How can we be moving that fast without noticing? How does it keep moving? And how can, say, water's motion be directed towards its proper place in the universe (above earth and below air) if that place is hurtling through space? Copernicus and Kepler want to tell us *how* the planets move, but at the expense of understanding *why* they move, or indeed why anything moves.

The universe of Kepler is a major Aristotelian headache!

Thus, Copernicus's revolution, even with Kepler's improvements, could not stay confined to astronomy. Physics had to be rewritten. While René Descartes (1596–1650) had championed the importance of expressing physical laws in the language of mathematics, and Galileo Galilei provided some crucial observations of gravity and motion, it was Isaac Newton who completed the revolution.

Newton's Universe

Newton proposed three simple laws of motion. His goal was not merely to shift the Sun, but to replace Aristotle's universe from top to bottom.

The key is *Newton's first law*. In 1687, he wrote (in Latin):[8]

Every body perseveres in its state of being at rest or of moving uniformly straight forward, except insofar as it is compelled to change its state by force impressed.

In other words, steady motion in a straight line does not need a constant push. A puck sliding on ice will keep sliding on its own. *Changes* in motion require a push or a pull.

Applying this to the Solar System, the planets don't need to be *pushed around*, that is, pushed from behind. But they do need to be *pulled in*, since otherwise they'd continue on a straight line off into space. That brings us to *Newton's second law*: an object changes its motion depending on the imposed force (push or pull), and on the mass of the object.

For the planets, that pull is provided by gravity: every object in the universe exerts a force on every other object in the universe, and the more mass, the stronger the pull. The Sun's huge mass keeps the planets in their orbits, and the Earth's pull keeps the Moon in orbit around us.

Newton's third law tells us that if object A pushes or pulls on object B, then object B gives an *equal and opposite* push or pull to object A. This principle implies that an object as a whole cannot accelerate itself. Suppose we're off on an interstellar picnic, cruising through space. Having stopped on the way at a popular tourist lookout over a nebula, we discover that the engines have stalled. There's a mechanic just a bit further on – can we get there somehow? We come up with a plan: we all gather at the front of the spaceship, floating in zero gravity. We kick against the inside wall, sending the spaceship on towards the mechanic. Because of Newton's law, the wall will exert an equal and opposite force on us, so we fly towards the back of the spaceship. The overall result is that the entire spaceship-and-crew system has no net acceleration. We discover this when we collide with the back wall of the ship, which brings everything to a complete stop. The ship has in fact moved slightly forwards, but it isn't moving. To repeat the process, we'd have to return to the front of the ship. But this would end up shifting the ship slightly in reverse, back to where we started. This kind of "pulling ourselves by our bootstraps" just doesn't work.

That might sound disappointing, but it's great for astronomers. The internal pushes and pulls of a planet – its core, mantle, crust, and continents – do very little to its orbit. When we model the Solar System, we can consider the planets to be simple spheres of matter. The details don't much matter, so the planets move on simple, smooth, geometric paths. As Kepler showed from observations and Newton showed from his equations, the planets move on ellipses.

Together, Newton's simple laws of motion and his equation of gravity explain an extraordinary amount about the physics of the Earth and the heavens. This is crucial, because while the insights of Copernicus, Galileo, and Kepler were important, they raised as many questions as they answered. Aristotle's physics was a comprehensive, integrated system, and the revolution in physics could only be completed by a similarly comprehensive theory.

You can't just criticize a paradigm – you have to replace it.

Spreading out the Heavens

Returning to the topic of this chapter, we turn our focus to an often-overlooked aspect of Newton's revolution. In Aristotle's universe, the Earth does not spin, so the apparent revolution of the heavens is real. Figure 2.4 shows the paths of the stars near the celestial South Pole, traced as the shutter on a camera is left open for a few hours; in the foreground is the Anglo-Australian Telescope at the Siding Spring Observatory, on the edge of the

Figure 2.4. By leaving the shutter on a camera open for a few hours, the paths of the stars near the celestial South Pole are traced. In the foreground is the Anglo-Australian Telescope at the Siding Spring Observatory, on the edge of the Warrumbungle National Park. (Used with permission: © Australian Astronomical Observatory and David Malin.)

Warrumbungle National Park – you should visit, it's beautiful! For Aristotle, the simplest explanation was to affix all the stars to a single spinning sphere.

For Newton, by contrast, the Earth rotates while the stars stay still. Hence, there is no need to fix them all at the same distance. The stars could be spread out through space, which helps explains why some appear brighter than others.

However, Newton's revolution creates a new problem, one that did not bother Aristotle. If gravity means that everything in the universe is pulling on everything else in the universe, then how can all those stars just sit there? Why aren't they all rushing towards the centre of the universe, destined to collide into one giant, collapsing mass?

Newton's solution to this problem is ingenious. If the universe is infinitely large, and filled everywhere with stars, then *there is no centre*. The stars do not collapse because there is no centre for them to collapse towards. Pulled equally on all sides, each star in the universe remains in place.[9]

This, however, creates yet another problem, one that even Newton didn't see.

Seeing the Stars

We don't see an infinite number of stars in the sky, of course. The further away a star is, the harder it is to see. Past a certain distance, an individual star is invisibly faint. So, you might think, almost all of Newton's infinite stars will dissolve into darkness, leaving just a few thousand local stars to stand out against the night sky. However, this isn't right.

Let's take a bird's eye view of Newton's universe. In Figure 2.5, you are on the left, looking up into the universe. Past a certain point, as we said above, you cannot pick out individual stars. But, the further out you look, the more stars there are in your eyeline. You can't see any individual star, but you can still see the combined light of all the stars. Like a floodlight made up of many smaller bulbs, the distant universe can shine very brightly.

Figure 2.5. Looking into the distant universe, slice by slice. The further away we look, the smaller each star looks but the more stars we see in each slice. These effects compensate for each other, so each slice sends the same amount of energy to the telescope.

Does this combined light fade away as we view more distant parts of the universe? No, it doesn't. Consider Figure 2.5 again, in which we have broken up our view into the universe into strips of equal height. As we move further away, each individual star looks smaller, but there are more stars. Further away, we see a smoother glow of light as the individual stars start to blend together, but the combined effect is the same: each strip sends the same amount of light down to us.

But in an infinite universe, there are an infinite number of these strips. That doesn't mean that the sky is infinitely bright – the stars in front block out the light from the stars behind. What happens is that each additional layer of stars fills in the gaps in the night sky, until the whole sky shines like a star!

Think of it like this; if there are an infinite number of stars scattered through an infinite universe, and we can see them all and they are just sitting there, then no matter what direction you look in the night sky, your line of sight will eventually hit a star. Everywhere you look, then, should be as bright as the surface of a star. The entire night sky should be ablaze.

But the sky, at night, is dark. So, where did we go wrong?

This particular puzzle is known as *Olbers's Paradox*, after the nineteenth century astronomer Heinrich Wilhelm Olbers, though others discussed the problem before him.[10]

A number of failed solutions need to be discarded, so that we can see the problem more clearly. What if the universe were filled with clouds of dust that obscure the light from distant stars? The problem is that stars shine because their gas has been heated to very high temperatures. The visible surface of the Sun, for example, glows at 5,500 °C. Dust clouds between the stars would absorb starlight, getting hotter and hotter. Eventually, they too would be heated to the temperature of the star and begin to glow. The dust clouds would become part of the problem.

Figure 2.5 shows stars scattered randomly throughout the universe. But stars are, in fact, grouped into galaxies. Could the darkness of the night sky result from the gaps between the galaxies? This won't work either, because we can simply repeat our argument above using galaxies. Looking out to infinite distance, the galaxies will fill in the gaps in the night sky. Every line of sight out into the universe will eventually alight on a galaxy, and a star in a galaxy. To escape Olbers's paradox in this way, we would need the infinite number of stars in the distant universe to be hiding *exactly* behind the finite number of stars we see in the night sky. And that seems implausible.

More generally, the distribution of stars in the universe as a whole would need to be a *fractal* pattern to avoid Olbers's paradox; that is, it would need to have grouping and clustering on every scale: stars in galaxies, galaxies in clusters of galaxies, clusters in super-clusters and so on forever. This pattern would also need to be quite sparse, which we don't see in our universe.[11] There is no evidence for this kind of fractal distribution in our universe – at scales larger than superclusters, the universe looks uniform.

What do we do with a failed theory? A scientific theory is a set of assumptions about how the universe works; science happens when we check to see whether we were right. If we find out that we were wrong, we must go back to the theory and change something. What, exactly, were the assumptions of the theory, and which, if changed, would make it correct?

So, what did we assume to reach the (incorrect) conclusion that the night sky should be as bright as the Sun? We assumed that:

A. The universe is infinitely large;
B. The universe is filled everywhere with stars;
C. We can see infinitely far into the universe;
D. Stars in the distant universe look the same as stars in the local universe.

Let's consider each of these in turn – can we make an alternative assumption and escape the false conclusion?

We could deny A and B, supposing that there are only a finite number of stars and nothing beyond. As in the universe of the ancients, there simply aren't enough stars to fill in the entire night sky, and where there is darkness, there are no stars. The problem, as we saw above, is that gravity is pulling on everything in the universe, so this finite system of stars would collapse.

As an aside, we cannot simply deny A. We tend to think of space as infinitely large and extending in all directions. This space is called *Euclidean*, after the Greek mathematician Euclid, and obeys all the familiar geometry you learned in high school: parallel lines never meet, the internal angles of a triangle add up to 180°, etc. But mathematicians have discovered that there are other ways that space could be. For example, space could be a three-dimensional version of the surface of a sphere. Think of the surface of the Earth. There is no edge – you can't drive off the Earth – but it isn't infinitely large. It joins up with itself, so that if you keep travelling in the same direction, you eventually get back to where you started. The surface of the Earth is two-dimensional, but you can do a similar thing in three dimensions. And since we are talking about space itself, we don't need to think of this *three-sphere* as being embedded in a higher-dimensional space, as the surface of the Earth is embedded in our three-dimensional universe. (We'll discuss these ideas further in Chapter 8.)

But there is still a problem. If we lived in a finite, spherical universe, and B, C, and D were all true, then we could see around the universe. You could look at the back of your own head. Like looking in a room of mirrors, you could see the same star more than once. And so Olbers's paradox returns: wherever you look, you will eventually see a star, either directly or after the starlight had made a few trips around the universe. The night sky would still look as bright as the Sun.

What about C? The key here is *the speed of light*. Our eyes gather light from our surroundings, and from this light we form a mental picture of the world. But this light does not instantly reach us, since light travels at a finite speed. Because this speed is so enormous – a billion kilometres per hour – we don't notice it. You are not really seeing your tennis opponent across the net as they are *now*, but as they were about a tenth of a microsecond ago; that is, about one 10,000,000th of a second ago. But that's no excuse for missing your return of serve.

On astronomical scales, however, light has an awfully long way to travel. When we look at a star in the night sky, we see light that was emitted by the star a long time ago. If you are looking at Alpha Crucis, the bottom star of the Southern Cross, 3 quadrillion kilometres away (3,000,000,000,000,000 km), you are actually seeing the star as it was 320 years ago.

So, as we look out into the universe, we are looking back in time. This means that Assumption C, that we can see infinitely far into the universe, is really assuming that the universe is infinitely *old*. We need all that time for starlight to travel to us.

If the universe were only finitely old, then we would only be able to see a finite distance out into the universe. This could solve Olbers's paradox – there are infinite stars, but we cannot see all of them because light has travelled so far and no further. In the dark parts of the night sky, we can't see the stars … yet.

Finally, it may not be clear where we assumed D – stars in the distant universe look the same as stars in the local universe – in creating Olbers's paradox. The assumption is that the light from stars just travels to us – nothing happens to it along the way. We saw above that pervasive dust clouds are an example of an attempt to deny Assumption D, but one that fails.

There is another way to deny D. You may not have heard of the Doppler effect, but you have experienced it. When an ambulance drives past, you hear the siren drop in pitch. As the ambulance approaches you, its sound waves bunch up, which your ears register as a higher note. As the ambulance passes you and moves away, the sound waves spread out, which gives a lower note. The note that you hear depends both on the note that the ambulance plays *and* the relative speed of the ambulance with respect to you.

It's the same with light waves. If an ambulance were speeding past at close to the speed of light, then its flashing light would look bluer (bunched up, more energetic light) as it came towards you and redder (spread out, less energetic light) as it raced away.

So, if a star in the distant universe were speeding away from us, it would look redder than nearby, stationary stars. The faster it moved, the more its light would shift towards longer and longer waves. Remember that our eyes can only see a narrow range of light waves, specifically those between 380 and 740 billionths of a metre. Longer or shorter waves – x-rays, infrared from your TV remote, or microwaves produced in your kitchen – are invisible. So, a fast-moving distant star would appear black to the human eye.

There are other ways to shift the wavelength of light, but the effect is the same. We could solve Olbers's paradox if we could arrange for light from stars in the distant universe to be shifted so much that they appear dark. Then, when we see darkness at night, we are really looking at shifted – and thus invisible to our eyes – starlight.

How the Big Bang Theory Solves Olbers's Paradox

Recall our advice in Chapter 1: know your enemy. In each chapter, we will explain how the big bang theory deals with the evidence.

So, in a big bang universe, why is the sky dark at night? Two of the incriminating assumptions are false. Firstly, our universe has a beginning, and only began forming stars a finite time ago. The matter in our universe started out smooth, hot, and dense; as it expanded and cooled, small lumps and bumps were drawn together by the attraction of gravity. Within this expanding space, gravity pulls matter into large haloes, and then into spiralling galaxies, within which stars form.

The first star ignited around 13 billion years ago, a few hundred million years after the beginning of the universe. This does not mean that we can see 13 billion light years out into the universe, because in the time since the star emitted its light, it has been carried further away by the expansion of space. It is now about 45 billion light years away. But the important point is that we cannot see forever. In the *observable universe*, where light has had time to reach us, there aren't enough stars to cover the sky and make it shine like the Sun.

Secondly, because the universe is expanding, light from the furthest reaches of the universe is shifted to longer wavelengths. Even if we could see forever, stars in the distant universe would be invisible to our eyes. We'll say more about this in the next chapter.

When we see darkness between the stars of our night sky, we are seeing back to a time before there were stars, and to a time when the expansion of the universe would have made any stars that were present invisible to our eyes today. The next time you admire a sparkling night sky, take some time to appreciate the darker hues of the cosmic palette.

A Rival Falls: A Static, Eternal, Infinite Universe

> ... constant stars, in them I read such art
> As truth and beauty shall together thrive.
>
> *Sonnet 14*, William Shakespeare

Olbers's paradox doesn't tell us what the universe is like. Observing that the night sky is dark between the stars doesn't answer all our questions about what is out there, what it is doing, and where it came from.

But it does rule out an otherwise live possibility. The universe cannot just sit there from eternity, filled with stars from end to end. *Something* in that picture must change, because it makes an undeniably wrong prediction – that the entire night sky would be as bright as the surface of the Sun.

There is an important lesson to be learned from Olbers's paradox – explore the consequences of your idea. *All the consequences.* If you have an idea (or a theory, or a grand principle, or a guess) about how the universe works, then you cannot cherry-pick the nice bits and dismiss the rest. You can't just say, "Well, *of course* the sky is dark at night. Everyone knows that." Your theory has a life of its own – if *it* doesn't know that the sky is dark at night, then it's wrong.

This is the job of the theoretical physicist, revolutionary or not. Ideas have consequences; you need to explore *all* of them.

3 RUN FOR THE HILLS!

Of the assumptions that create Olbers's paradox, two involve infinities. Is the universe infinitely large, or infinitely old, or both? It's hard to know where to start with such questions.

The last assumption is easier: is the universe static? Is it just sitting there?

For thousands of years, the answer was obvious. The heavens were a paradigm of perfection: the constant light of the Sun, the flawless circular paths of the stars, the planets wandering night after night. Sure, the occasional comet or eclipse upsets the order of the cosmos for a while, but things soon return to business as usual. Nature moves in cycles, but nothing much changes from cycle to cycle.

Even with a powerful telescope, you won't see much of a change in the universe from night to night. But there are other ways to see if the universe is moving.

Red Light, Blue Light

In the last chapter, we introduced the Doppler effect for light: if a star is moving away from you, its light will be stretched to longer wavelengths, towards the red end of the spectrum, and if it is moving towards you, light will be shifted towards the blue. The same is true for the combined light of all the stars in a distant galaxy.

Can we use this effect to measure how fast stars and galaxies are moving? There seems to be a problem with this idea. To know whether a star or galaxy is moving towards or away from us, we need to know whether its light is redder or bluer than it *would* look if were sitting still. But how could we know that? How do we know whether a distant galaxy looks red because it's speeding away from us, or because it's just a red galaxy?

Consider the case of sound. The pitch of a sound wave is directly analogous to the colour of light – both depend on the length of the wave, from crest to crest. Suppose we hear the melodious tune of an ice-cream van in the distance. If we knew the pitch of the tune that the van was playing, we could measure the pitch that we hear and work out whether the van was moving towards or away from us. In fact, we could work out how fast it was moving towards or away, based on how much the tune was shifted.

For example, if the van plays *Greensleeves* in the key of C, but we hear the tune one semitone higher (in the key of C#), then from the definition of the musical scale, we know that the pitch of the song has shifted upwards by 6%. So, we know that the van is moving towards us at 6% of the speed of sound in air: 73 km/h. (Sound travels at 1,235 km/h in air.)

However, there is no standard recording of *Greensleeves* to which we can compare. If we can't see the van, then for all we know it could be stationary and playing the tune in the key of C#. Or playing the tune in the key of D, and moving *away* from us at 73 km/h. Or it could be a drag racer, playing *Greensleeves* in G while dispensing ice cream at 500 km/h.

But suppose our friendly neighbourhood ice-cream vendor grows tired of the same old tune and reaches for a classic: Jimi Hendrix's *Voodoo Child (Slight Return)*, live at Woodstock. Sitting in our lounge rooms, if we recognize the tune, we can get our own copy of Hendrix out of our music library. Now we have the extra information we need – we have the same sound that the van is playing, but without the pitch-shifting effects of speed. We can compare the pitch of the two sounds and work out the direction and speed of the van. Hendrix famously detuned his guitar by a semitone, so the recording is in E flat. If the sound we hear from the van is in E (natural), then the van is coming towards us at 73 km/h.

Let's return to light. Is there a substance in the universe that emits a recognizable light "tune"? Can we scan the heavens for the mythical substance *hendrixium*, which glows with a purple haze?

The answer is *yes!* It's everything astronomers could want (excuse me, while I kiss the sky!). It's the key to deciphering the motions of the heavens – we can work out if it's coming up or down.

The Music of Atoms

It's atoms. Every type of atom (or *element*) in the periodic table – from hydrogen to uranium – can send us a specific light "chord", a collection of wavelengths of light that *only* that element emits.

Here's what it looks like. Our eyes only observe three wavelength ranges, which our brain interprets as the three primary colours. These three colours are hard-wired – there are three kinds of colour receptors, called *cones*, in our eyes. If we want more details about how much of each wavelength of light is in a given beam, we need to split the light into its different wavelengths. The most familiar way to do this is with a prism; think of the classic album cover for Pink Floyd's *Dark Side of the Moon* – white light goes in, and a rainbow comes out. We can shine the rainbow on a screen, so that the different wavelengths of light, from red to violet, shine in different places. If we shone a pure green light through a prism, then we would see a single bright green band on the screen, surrounded by darkness where the other colours are absent.

Here's where the magic happens. If we put some hydrogen in a tube and apply a large electric voltage, the tube will glow. Shine that light at a prism, and you will see an image like the one shown in Figure 3.1.

Fill the tube with helium, make it glow, and point it at a prism. What you will see is shown in Figure 3.2.

We can do this for all the elements, compiling the Periodic Table of Spectra.[1] Just pump enough energy into a cloud of

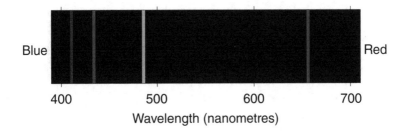

Figure 3.1. The emission spectrum of hydrogen. If you put some hydrogen in a tube and apply a large electric voltage, the tube will glow. Shine that glow at a prism, and this is what you see. Instead of the full rainbow of light, we see discrete bands of colour.

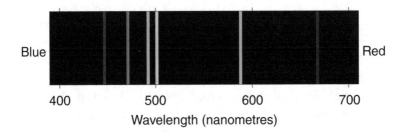

Figure 3.2. The emission spectrum of helium. Grab some helium, make it glow, shine that glow at a prism, and this is what you see. As with hydrogen, and indeed all the elements, we see discrete bands of colour.

atoms, via heat or electricity, and they will play their chord of light for you.

There is another way to see the characteristic chord of atoms. Just as atoms are more likely to *emit* light at certain wavelengths, they are also more likely to *absorb* light at those same wavelengths. So, a cloud of atoms will act like a fog, where the wavelengths of the chord won't make it through the fog. We then see an *absorption spectrum*, as shown in Figure 3.3.

To an astrophysicist, this is all utterly wonderful. With a device called a *spectrograph*, which (like a prism) measures the amount of light at a large number of different wavelengths, we can work out what the universe is made of. Each element imprints itself on our observations. In fact, this is how helium

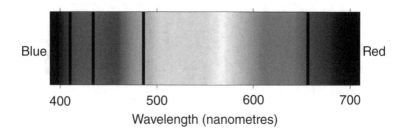

Figure 3.3. The absorption spectrum of hydrogen. If we shine white light at a cloud of hydrogen, and then through a prism, we find that specific wavelengths of light are missing, scattered by the gas. These lines correspond to the emission lines of hydrogen (Figure 3.1).

was discovered. In 1868, Pierre Janssen observed an unfamiliar spectral line in the light of the Sun and named the new element after the Greek god of the Sun, *Helios*. It was only much later, when helium was discovered on Earth and its spectrum measured in a laboratory, that chemists believed that the astronomers had really discovered a new element in the Sun.

So, by comparing our observations of the heavens with careful experiments on atoms in laboratories here on Earth, we can identify the signatures of different atoms. In particular, we can work out whether the light from the distant star or galaxy is shifted to shorter or longer wavelengths.

We need to pause for a moment. Until now, we have presented the shifting of light to longer or shorter wavelengths as being the result of motion (the Doppler shift). Shifting to redder, longer wavelength light means motion away; shifting to bluer, shorter wavelength light means motion towards. But what we observe is the shift. We *infer* the motion. We'll look at other explanations for the shift in later sections.

The important thing about these shifts is that they affect all wavelengths in the same way. We can make white light look red by passing it through tinted glass, for example. But this doesn't shift the light; it just blocks out some of the blue light. And remember that light comes in many more wavelengths than the red-to-blue colours that our eyes can see. There are radio waves and microwaves with much longer wavelengths, and ultraviolet light and x-rays with much shorter wavelengths. Radio waves and x-rays will pass through tinted glass unaffected.

By contrast, the Doppler shift affects all wavelengths of light in the same way. If blue light is shifted by motion to be double the wavelength, then so will ultraviolet light and radio waves. A one-nanometre wave (a billionth of a metre) becomes a two-nanometre wave, and a one-metre wave becomes a two-metre wave. Let's call this a *uniform shift*. We can check whether we see a uniform shift by comparing the shift of a number of different atoms, or different lines from the same atom.

Let's return to our spectrographs, pointing upwards to the night sky, ready to observe the light from the distant universe. What do we see?

Great Galloping Galaxies!

To a sufficiently powerful telescope, the night sky is not only full of stars, but galaxies of stars. We see nearby stars in our Milky Way galaxy as points of light, and we see the combined light of hundreds of billions of stars in distant galaxies as a smooth ball or disk of light.

When we point a spectrograph at a galaxy, we see the combined light from all the stars in the galaxy, as well as any other gas between the stars that is emitting or absorbing light. The same chords of light are visible, so we can determine what the galaxy is made of.

As with the ice-cream vending Hendrix fan, we can also compare the light from distant galaxies to the light from elements here on Earth, and see whether we see a shift, whether it is a *uniform* shift, and whether it is a *blueshift* (shorter wavelength) or a *redshift* (longer wavelength).

We have learnt a lot about the universe by surveying galaxies with powerful telescopes. In the Northern Hemisphere, the *Sloan Digital Sky Survey* (SDSS), the largest survey to date, has measured more than a million individual galaxies over a third of the sky. In the south, the *2dF Galaxy Redshift Survey* (2dFGRS) has mapped

250,000 galaxies with the 3.9-m Anglo-Australian Telescope located in Australia (as seen in Figure 2.4).

These surveys see redshifts galore. Out of over a million known galaxies, only about 50 show blueshifted light. These blueshifts are quite small, about 1% at most. For the vast majority of galaxies, their light is redshifted to longer wavelengths, and by significant amounts. If these redshifts are due to the Doppler effect, then the vast majority of galaxies are moving away from us at enormous speeds, close to the speed of light.

How Far to the Stars?

Why are (almost) all galaxies redshifted? And by such large amounts! It might help to know how galaxies are arranged in the universe. Can we map the galaxies?

To do this, we need to be able to measure the distance to a galaxy. Obviously, this is no easy task. We can't go there, measuring off each kilometre that we travel. We can't send out a tape measure. We can't bounce light off it and measure the amount of time it takes to get back.

But there is a way of measuring distances to far off places that doesn't involve travelling there. The idea is shown in Figure 3.4. You start with a single baseline, accurately measured. Then, you need a theodolite, which is a surveying instrument used to

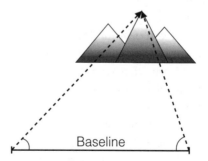

Figure 3.4. Measuring the distance to a far-off mountain, using a baseline and two measured angles.

measure *angles*. Now, suppose you want to measure the distance to a mountain. Position your theodolite on one end of the baseline, and accurately measure the angle between the baseline and the mountain. Repeat for the other end of the baseline. With a bit of high-school geometry, you can use the two measured angles and the length of the baseline to calculate the distance to the mountain.

This assumes that your baseline is long enough, and your theodolite is accurate enough, to see the difference in the angle. If your baseline is only a few metres long, and you measure angles by eyeballing your old high-school protractor, then don't expect to survey your local mountain range.

But the technique can be extremely useful and accurate. In the nineteenth century, before the advent of modern infrared distance measurement, India was surveyed beginning near Chennai. The baseline was a 12-km steel chain, supported by wooden boxes, and covered along its entire length by tents to minimize its expansion under the heat of the Indian sun. The surveyors even knew that the chain would expand by 0.3 millimetres if the temperature went up by 1 °C. Accuracy of the baseline was crucial, of course: by the time they had used a series of triangles to measure their way across India to another steel baseline near Bangalore, 260 km away, they were accurate to within 10 cm.

Archimedes once said, "Give me a place to stand and with a lever I will move the whole world." Can we find *two* places to stand, with an accurately measured distance between them, from which to reach for the stars?

Remember that the distance we can measure depends on the length of our baseline, and the accuracy with which we can measure angles. Given the accuracy of even the most advanced space satellites available today, the Earth, unfortunately, is not enough. If we send our satellite from one side of the Earth to the other and try to measure the distances to the stars, we will be successful for about 50 of our nearest stellar neighbours. The rest are simply too far away for our instruments to measure any difference in their position from the ends of the baseline. It's a

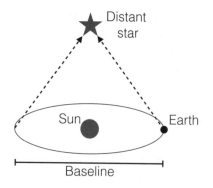

Figure 3.5. Measuring distances in the universe using the same principle as in Figure 3.4 – a baseline and two measured angles.

start, but hardly enough to bridge the gap to other galaxies. We need a wider baseline.

And nature, graciously, has provided one; the Earth moves around the Sun. Instead of sending our satellite to either side of the Earth, we wait six months for the Earth to be on the other side of the Sun. Now we're in business: with this 300-million-kilometre baseline, we can measure the distances to stars that are tens of thousands of light years away (see Figure 3.5).

Astronomers call this technique *parallax*. We measure the change in position of a nearby star, relative to the more stationary background of far-away stars. This is analogous to the way nearby objects fly past the window of your car, while far away objects pass by more slowly, and the Moon seems to follow you. The more a star appears to move between January and July, the closer it is.

Still, we can only measure the parallax of stars in our Milky Way. We can't measure the distance to other galaxies, as the difference in angles is far too small to measure. But we've made a start.

Blinking in the Dark

The width of the Earth's orbit is the longest baseline we can hope for in our corner of the universe. Sure, the Solar System is orbiting around the centre of the Milky Way, but we can't really afford to wait a hundred million years to get to the other side.

We're going to need something at the other end. If we had a known baseline out there in the universe, and could measure the angle between its extremities, then we would be able to calculate its distance. This flips Figure 3.5 upside down. How big something *appears* depends on how big it is, and on how far away it is. A baseline of known size out in the universe is called a *standard ruler*.

How else will the appearance of an object depend on the distance from which it is viewed? The more distant a light source is, the *dimmer* it appears. We saw this in our discussion of Olbers's paradox. A lamp, when viewed from twice as far away, appears four times dimmer. So, if we had an object whose intrinsic brightness was known, then we could measure how bright it appears, and from this calculate how far away it is. Such objects are called *standard candles*.

So, has the universe been kind enough to provide us with standard rulers or standard candles? Not directly. We can't see galaxies out there with giant tags that read:

Question 1: I'm 100,000 light years from one side to another. How far away am I? (5 Marks)

If we're going to use a standard ruler or standard candle, we need to be able to infer its intrinsic size or brightness from a property *other than* its apparent size or brightness. Let's call this property *the label*.

Suppose you think you've got a label: you discover a rosy-pink star, and your best theory about how a star can be that colour says that these "Rosie stars" are all exactly as intrinsically bright as the Sun. Great! The label is *pinkness*: wherever we see a Rosie star, all we need to do is measure how bright it appears and we can infer how far away it is. And if the universe has a whole lot of Rosie stars, we can measure the distance even if it's too far away for parallax to work.

But wait – how can we be sure that all Rosie stars are exactly as bright as the Sun? They could all be half as bright as the Sun, and closer than we think. Here's where parallax comes into play. It can't reach the galaxies, but if it can reach a sample of nearby Rosie stars, then we can calculate the distance to each one. From their distance and apparent brightness, we can calculate their intrinsic brightness. And from this, we can check whether they're all the same. This would also work if you discovered a population of dark green galaxies that were all the same size. These "Jade Exact Diameter" or JED galaxies would be perfect standard rulers.

The actual universe provides us not with Rosie stars or JED galaxies, but with *Cepheid variable stars*. They aren't pink; they brighten and fade with remarkable regularity. The star is oscillating, contracting, and expanding under the effect of gravity pulling inwards and the internal pressure of the star pushing outwards.

At this point, meet the hero of our story: Henrietta Leavitt, an astronomer working at the Harvard College Observatory in Massachusetts at the start of the twentieth century. She was studying stars in the Small Magellanic Cloud (SMC), a dwarf galaxy in the neighbourhood of our Milky Way. Because the SMC is a compact collection of stars, the stars are approximately at the same distance from us. Leavitt noted that the stars that *appeared* brighter took longer to fade and return to full brightness. She concluded that the same was true of stars that *actually were* brighter. She was quick to note the usefulness of these variable stars: "It is to be hoped, also, that the parallaxes of some variable stars of this type may be measured," she wrote in 1912.

So, we have a label! We have since measured the parallaxes to a host of local Cepheid variable stars. So, we know the relationship between the time it takes for a Cepheid to fade and brighten, and the intrinsic brightness of the star. So, if we spot a blinking star in the distant universe, even if we can't measure its parallax, we can calculate its distance. Crucially, Cepheids are bright enough to be spotted in distant galaxies – galaxies with redshifts!

But we can do even better.

Far Away Fireworks

While we can see Cepheid variable stars in other galaxies, they are still only single stars. Even the Hubble Space Telescope's key project to measure distances to galaxies has only spotted 23 Cepheids in nearby galaxies. The rest of the millions of galaxies we can see in the night sky are too far away for us to spot a single blinking star.

Can we find another label, another property of *something* out in the universe that tells us how intrinsically bright it is? And we'd like it to be brighter than a single star, even a big blinking one.

And once again, nature has provided. What's brighter than a single star, you ask? A star that blows up! In stars, there is a balance between gravity pulling the mass inwards, and pressure keeping the star up. Usually, these two keep each other at bay – a bit too much pull from gravity, and the pressure will fight back. A bit too much pressure pushing out, and gravity will draw everything back in.

But in certain circumstances, the pressure inside a star loses its composure, over-reacts and blows the star to smithereens. This is called a supernova. For example, when a very large star runs out of useable fuel and collapses, the outer parts of the star pile up against the inner, burned-out core. This traffic jam creates a massive excess of pressure, and the star rebounds, blasting matter into the universe.

There is a specific type of supernova that is particularly useful for our purposes. It involves an object called a white dwarf, which is a very compact ball of matter that remains after a small star has burned itself out. It is not big enough and hot enough to squeeze its insides and ignite more nuclear reactions, so it just sits there, cooling off.

But the universe is a busy place, and some white dwarfs don't enjoy a peaceful retirement. They have a nearby companion – a star or a planet or another white dwarf. If the companion is close enough, it will feed the white dwarf, dropping matter onto its surface. As the mass of the white dwarf grows, it will squeeze its insides and raise the temperature.

In some white dwarfs, a line will be crossed. Very suddenly, the additional squeeze will reignite nuclear reactions. Rather than the careful balance between gravity pulling in and pressure pushing out that we find in normal stars, the white dwarf will ignite a very dense, very hot interior. The fuse is lit, an enormous amount of pressure is generated, and the white dwarf is obliterated. For a short period of time, the explosion is billions of times brighter than a star.

Because the process that ignites the explosion involves a very specific threshold, these *Type 1a supernovae* are all very similar. They are a very good standard candle, but we can do better. In 1993, astronomer Mark Philips showed that there is a neat relationship between how intrinsically bright a Type 1a is and how quickly the light from the explosion fades. We've found ourselves another label, this time on supernovae explosions rather than stars.

But how can we test the label? How can we make sure we put the right label on the right Type 1a supernovae? As before, we need another way to measure their distance. But we don't have a Type 1a close enough for parallax to work, which is a good thing. Supernovae are a tad nasty for nearby life forms.

But we have another distance measure – Cepheids. We know that the Cepheid label measures distance, thanks to parallax. So, we can use galaxies in which we find both a Cepheid and a Type 1a supernova to check our new label.

This is called the *distance ladder*. Each rung – parallax to Cepheids to Type 1a supernovae – sends us further out into the universe, and rests firmly on the rung below.

And the Type 1a rung is *long*. To illustrate, the nearest large galaxy to our Milky Way is Andromeda, located about 2.5 million light years away. That is, light takes 2.5 million years to get from Andromeda to Earth. With Cepheid variable stars, we can reach out 75 million light years, or 30 times further than Andromeda. With Type 1a supernovae, we can reach out 10 billion light years, or about four thousand times further than Andromeda. Recall that the universe is only about 13.8 billion years old, so we are reaching a fantastically long way across the universe with these remarkably useful explosions.

Finally, we can do it! Using the distance ladder, we can measure the distances of redshifted galaxies across the known universe. So, what's going on out there? How are they arranged?

Redder and Further

In 1929, Edwin Hubble was the first to put enough pieces together to give us a view of the universe at large. He had the velocities of distant galaxies moving away from us, observed and calculated by the astronomers Vesto Slipher and Milton Humason. He had made a study of Cepheid variable stars in galaxies and, thanks to the work of Henrietta Leavitt and others, could convert these observations into distances. He didn't have supernovae – those would come 60 years later – but he had enough distances from Cepheids to be able to map the nearby universe and its motion.[2] His results are shown in Figure 3.6.

Even with this limited and scattered data, we can see a trend between the two quantities: galaxies that are further away

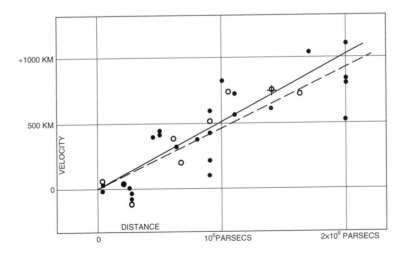

Figure 3.6. Hubble's 1929 data,[3] plotted as distance on the horizontal axis, and redshift (expressed as a velocity) on the vertical axis. The proposed linear relationship is shown as the diagonal lines.

(towards the right of the plot) tend to be more redshifted (towards the top of the plot). Hubble's data is consistent with a linear relationship: if you look twice as far away from Earth, the galaxies there will be twice as redshifted. Three times away, three times as redshifted. And so on. The slope of this line is a number known as the *Hubble constant*.

Here, we're interested in the best data we have today, not just the data from 1929. You're too late to launch a revolution against Edwin Hubble! This is an important distinction because Hubble's data is limited, and the relationship he used to convert Cepheid observations to distances had confused two different types of blinking stars. Until this was sorted out, things were a bit confused.

But today, the equivalent diagram is well understood and well studied, and is shown in Figure 3.7.

We need to pause to explain something important about Figure 3.7. You will notice that we haven't plotted measurements as single points, but as crosses. These *error bars* represent the uncertainty of each measurement. Sources of uncertainty can creep into our data in many ways: sources of light other than the object we're looking at, limitations in our instrument, or the jittering of electrons in the wires of a detector. This noise cannot be completely eliminated, and so it must be reflected in our conclusions. Typically, scientists quote and plot "one-sigma" uncertainties: for example, if we report a measurement as 1.56 ± 0.04, we mean that we're 68% sure that the true value is between 1.52 and 1.60. In many cases, the uncertainty is spread out so that we are 95% sure that the true value is within "two sigma" (± 0.08), and 99.7% sure that the true value is within "three sigma" (± 0.12).

Figure 3.7 shows one-sigma error bars in the horizontal and vertical directions, indicating that there are uncertainties in both the measurement of distance and redshift. Each measurement, then, is not a single sharp point but more like a fuzzy oval shape, delimited by the error bars. A theoretical prediction is considered successful if it passes through most of these fuzzy ovals.

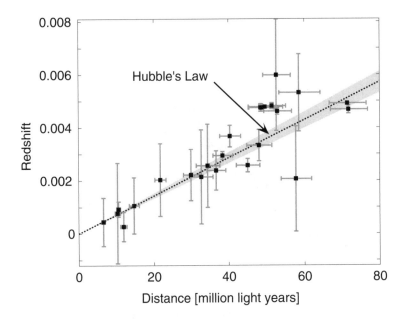

Figure 3.7. The modern-day Hubble diagram for Cepheid variable stars, relating distance to redshift. The grey bars on each point represent the uncertainty of each measurement. The dotted line and grey wedge show Hubble's Law – redshift is proportional to distance – for modern values of the Hubble constant. (This data, reported in 2001, was taken by a team led by Wendy Freedman that used the Hubble Space Telescope.[4])

Looking at the data, we clearly see Hubble's Law, that redshift is proportional to distance. But, as we saw above, the most important use of Cepheids is as a rung to reach Type 1a supernovae. Using these, we can plot the distances and redshifts of over a thousand supernovae at enormous distances (Figure 3.8).

Two separate scientific collaborations have collected this data, shown as squares and triangles. Vertical error bars are too small to plot, so only the horizontal error bars shown. The trend is undeniable: the further away a galaxy is, even tens of billions of light years away, the more redshifted its light. These are all uniform shifts.

So, we ask again, what is going on out there?

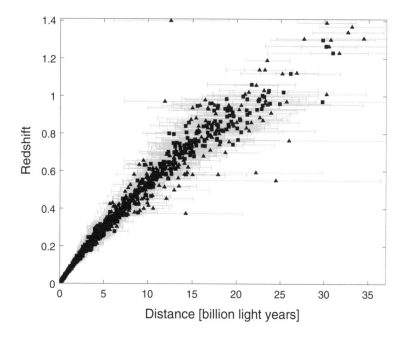

Figure 3.8. The Hubble diagram of distance vs. redshift, for observations of distant supernovae. The two different markers (squares and triangles) represent two independent scientific groups that have observed supernovae.[5] Note that the horizontal axis measures *billions* of light years.

How to Shift Light

It is worth stressing just how uniform these shifts are. The musical chords of atoms, which allow us to measure how much the light from the star or galaxy has been redshifted or blueshifted, can range from radio waves from hydrogen that are 21 cm from crest to crest, to x-rays from iron that measure less than a millionth of a millimetre. There are spectral lines from molecules that are similar to the microwaves in your kitchen, and even closely-spaced doublet lines from sodium and triplet lines from oxygen. We see many of the elements in the periodic table, and most of the molecules that make up a good beer, all imprinted in light thanks to the music of atoms.

The instruments we use to measure these spectral lines range from huge radio telescopes, which can be dishes or arrays of antennae, to optical telescopes with mirrors that are 10 metres in diameter, to x-ray satellites that require a delicate array of nested mirrors. And it doesn't matter which of these we use to measure the redshift of a distant galaxy. They all agree. Whatever is stretching the light from these galaxies doesn't care whether it's radio, microwave, visible, ultraviolet, or x-ray.

No known material can do this. That is, there is no known substance that will uniformly stretch or compress light waves of all sizes. The reason is quite simple. Suppose we had such a wonderful substance that could shift light uniformly; call it *Morphium*. Light is an electromagnetic wave, so to affect the passing light rays, we need Morphium to contain electrically charged parts. Like a wave on the ocean that causes a boat to rise and fall, light will cause matter to oscillate back and forth, absorbing some of the energy of the wave. Conversely, if electrically charged matter is wiggled back and forth, we create light. The faster the wiggle, the shorter the wavelength of the light.

So, to achieve a uniform shift, we need Morphium to absorb wiggles of all sizes, and re-emit this light stretched by a fixed amount. Morphium must be perfectly accommodating: its wiggles must be determined *only* by the light it absorbs. Any personality, any quirk, any internal property of Morphium that imprints on the wavelength of the re-emitted light will ruin the required uniform shift.

But that's the problem. Because the internal parts of Morphium are electrically charged – as they must be to affect light at all – they will interact with *each other*. For example, atoms are held together by electromagnetic forces. Negatively charged electrons orbit the positively charged nucleus, bonded by electric attraction. And atoms can be bonded to each other in fixed, rigid lattices by the combined pushes and pulls of all the other atoms and their charged protons and electrons in the grid. These lattices are what we call solid objects.

When light strikes matter, it does not find a perfectly flexible medium, waiting to respond to its every whim. It instead finds a highly-strung array of charged particles, tightly held in place.

For this reason, every known material affects different wavelengths of light in different ways. Some wavelengths just pass right through, some are modified and moulded by the lattice, and some just bounce off the surface.

But that's solid objects, you might say. What about liquids and gases? Liquids won't help, because while their atoms don't form a rigid lattice, they are still a hotbed of racy atom-on-atom and molecule-on-molecule interaction. For the same reason, a gas of molecules won't help either, since the electric forces between the atoms that make up each molecule will react to different wavelengths of light very differently. Even a gas of pure atoms – say, atomic hydrogen, with one electron orbiting a single proton – has too much personality to make a uniform light stretcher. And even if we just had electrons, the way that they wiggle depends on how heavy they are. Think of our boat on the ocean again – a kayaker will feel every ripple, whereas the passengers on a cruise liner remain blissfully unaware of all but the largest swells.

There just isn't a way to make a uniform light stretcher out of matter. Which doesn't leave us with many options. If it isn't the material of the universe that's stretching the light, then the only other option is space and time themselves. We've seen the simplest example of this already – the Doppler shift, which is caused by motion of the source of light through space.

If space and time were merely the static stage of cosmic action, then that would be the end of our story: uniform shifts are Doppler shifts. But Einstein taught us that space and time are dynamic things, able to stretch and bend. This *spacetime curvature* is gravity. We can think of the gravity of the Sun as being like a hole with sloping sides (Figure 3.9). The Earth rolls around in its orbit as if on a banked curve.

Because gravity affects space and time itself, light too will feel its effect. When light travels to us from the Sun, it must climb out of the hole. We can think of this as the light losing energy as it climbs out of the reach of the gravity of the star, resulting in less energetic, redder light. If we play this scenario in reverse, we see that light travelling downwards towards the Sun is shifted to

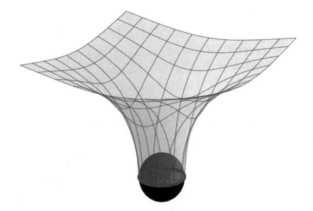

Figure 3.9. The picture of the curvature of spacetime, which is the cornerstone of the general theory of relativity. The central mass distorts the spacetime around it, making nearby objects move on apparently curved paths.

the blue. This is known as *gravitational redshift* (or blueshift). And because the effect is due to space and time themselves, it is a uniform shift.

This is not just theory. Experimenters first detected gravitational redshift in 1959. Two Harvard physicists, Robert Pound and Glen Rebka, fired photons (particles of light) of light up and down a tower.[6] The expected redshifting and blueshifting due to the change in the gravitational potential between the top and the bottom of the tower is tiny, of the order of 1 part in 10^{15}. Measuring such a minuscule shift was only possible due to a discovery by Rudolf Mössbauer the year before, namely that atoms held tightly in crystals do not rebound when a photon is emitted.[7] As a result, the wavelength of the emitted radiation is remarkably consistent, allowing the gravitational shift to be measured. The resulting shift was consistent with the prediction of Einstein's theory, to within 10%.

About the same time as Pound and Rebka were experimenting, astronomers were detecting the gravitational redshift of light climbing out of the deep potentials at the surface of stars. The first conclusive observations involved white dwarfs, the crushed cores of dead stars. More recently, astronomers have detected the gravitational redshifting due to the Sun.

Could the redshift of distant galaxies be due to gravitational redshifts? The idea is that, as we look further and further into the universe, galaxies create deeper and deeper holes for their stars.

But there's a problem. Einstein's theory tells us that the amount of redshift depends on how deep the hole is, which in turn depends on how much mass is squeezed into the hole. For example, the Earth squeezes 6 billion trillion tonnes of matter into a region that is 12,742 kilometres across. Einstein's theory tells us that light shone upwards from the surface of the Earth is redshifted by a tiny factor: less than one in a billion.

It's a start, but we need more! We'd get a larger redshift if we could compress the mass of the Earth into a smaller region. This deepens the hole and makes the outward climb more taxing. How much would we need to squeeze the Earth by to get the large redshifts we see from distant galaxies? A lot! We would need all the mass of the Earth to occupy a region a few *centimetres* across.

So, suppose we see a galaxy that has been redshifted by a factor of 100% – every wavelength of light has been stretched to double the size. Type 1a supernovae studies tell us that the galaxy is 25 billion light years away. That's an awfully long way, but with our best telescopes we have been able to observe millions of such galaxies.

We can use the amount of light emitted by the galaxy as a rough guide to how many stars it has, and ask: how tightly do we need to pack those stars for gravitational redshift *by the galaxy itself* to explain its redshift? Einstein tells us that all the hundreds of billions of stars in the galaxy would need to be orbiting a black hole, perilously close to its *event horizon*, that is, the point of no return. Given the number of stars that we can observe in a nearby galaxy such as Andromeda, we would need all this mass to be compressed into a tiny region: about 1% of the distance from the Sun to our nearest star.

And there's the problem – we can see the galaxy. We can see how big it appears, and because we know how far away it is, we can calculate that it's millions of times too big. Even if there were a black hole at the centre of the galaxy – and there's good evidence that most galaxies do have a black hole at their centre – the vast majority of the stars are too far away to be significantly gravitationally redshifted.

We don't see any of the other signatures of black holes, either. To avoid falling into the black hole, a nearby star would have to orbit at close to the speed of light. But then we would see Doppler shifts as well, as stars are violently flung around. About half the stars would be going away from us and half towards us at tremendous speeds. Indeed, we can measure the Doppler shifts of the stars in galaxies. They orbit the centre of their galaxies, but at speeds that are thousands of times too slow. This precludes the possibility that the redshift is caused by an enormous amount of unseen matter.

Let's review. We are trying to answer the question, why is the light from distant galaxies redshifted? We have whittled our way down to a small number of options. Given the uniformity of the shifts – that all wavelengths of light are stretched in the same way – electrically charged matter is not responsible. Gravitational redshifts by the galaxy itself aren't large enough, because galaxies are too spread out and rotating too slowly to be the deep gravitational wells we require. The only options remaining are Doppler shift, or a stretching of the light along the way from the galaxy to us, or both.

Here is the challenge for the cosmic revolutionary. Like recognizing a Hendrix guitar solo in a higher key, we can recognize chords of light from atoms in the distant universe. The light from the vast majority of galaxies is redshifted, that is, stretched to longer wavelengths. And stretched by a substantial amount – the current record holder is a galaxy whose light is *ten* times longer today than when it departed. Furthermore, we can measure distances to stars directly with parallax, and can use this information to infer the distance to blinking Cepheid variable stars. We can use Cepheids to infer the distance to Type 1a supernovae, and thereby measure the distances to over a thousand galaxies with a range of redshifts.

We find that the further away a galaxy is, the more its light is redshifted. The relationship, especially out to a few billion light years, is roughly linear – twice as far means twice as shifted. We see this no matter where we look in the sky, and no matter what telescope we look with. That's the data, known to cosmologists as *Hubble's Law*. Your revolutionary idea about the universe is going to have to deal with it.

How the Big Bang Theory Explains Hubble's Law

According to the big bang theory, space itself is expanding. Let's unpack that idea.

Consider two intrepid explorers, Lois and Clark, who are standing one kilometre apart on the Earth's equator. They get out their compasses, find due south, and begin walking. We'll suppose that the Earth is perfectly smooth, and they needn't worry about oceans or lakes. As they set off, their paths are parallel, perfectly perpendicular to the equator. And yet, as they journey, something strange happens. They're walking straight ahead, but they seem to be getting closer together. The effect is slow but undeniable, and by the time they reach the South Pole, they run into each other!

What happened? We know what really happened – they're walking on a sphere. Their paths started out parallel (perpendicular to a straight line between them), they haven't deviated to the left or right, and there's nothing pushing them together. But they're walking on the curved surface of Earth.

If, however, they thought that the Earth was flat, they must conclude that a force is pulling them together as they move. With a bit of experimentation, they discover something curious about this force. If Clark rides an elephant, they think, the force will have a bigger object to shift and so it will draw them together more gradually. But they are wrong: Lois and Clark run into each other at the South Pole again.

So, they surmise, the force drawing them together must get stronger as the objects in question get heavier. And, it must get stronger in just the right way for the mass of the object to be irrelevant to its path. This strange coincidence – a force that knows how to make the mass of the object irrelevant – is a clue that it's not a force at all. Lois and Clark are not following curved paths on a flat surface. They are following *locally* straight paths on a curved surface.

According to Isaac Newton, gravity has the suspicious ability to make the mass of the object irrelevant to how it moves. Like the push on the elephant, Newton's force of gravity gets stronger with the mass of the object being pushed. Einstein was the first to take the hint and formulate a *geometric* theory of gravity. Lois and Clark moved on a curved surface in space, but to explain gravity, Einstein supposed that space *and time*, considered as a four-dimensional entity called *spacetime*, is curved.

Obviously, we're glossing over a few mathematical details. But the key insight of Einstein is that space and time form a unified, dynamic entity, whose warping and curving influences how objects in the universe move. We tell Einstein's equations where the matter is and where it's moving, and Einstein's equations tell us how spacetime warps and curves.

We can use the distances we discussed in previous sections to make a map of the galaxies in the universe. A small piece of that map is shown in Figure 3.10, drawn from the millions of galaxies in the Sloan Digital Sky Survey. We see that, if we zoom out and look at the big picture, galaxies seem to be arranged fairly evenly in space. We have to be careful interpreting the data: it is easier to see galaxies that are nearby, which could fool us into thinking that there are more galaxies in our neighbourhood. Taking this into account, our observations of the universe suggest a simple overall arrangement – the universe is, on average, the same everywhere.

So, according to Einstein's theory of gravity, what happens to a universe that is uniformly filled with matter and energy? There are only two options: expand or contract. Any other kind of motion would distort the arrangement. And because the

Figure 3.10. A projected map of galaxies in the universe from the Sloan Digital Sky Survey (SDSS). The image shows 48,741 galaxies, about 3% of the full survey data set. This slice of the universe is 6 billion light years wide, 4.5 billion light years high, and 500 million light years thick. The galaxies are, on average, uniformly spread. (Used with permission: Daniel Eisenstein and the SDSS-III collaboration.)

matter and energy in the universe affect spacetime, the universe won't just sit still.

According to Einstein, the galaxies in the universe are not expanding into empty space; space itself is expanding. If the universe is finite in size, then the expansion of the universe is – quite literally – creating more space. You can measure its volume, and that number is getting bigger. You could fit more oranges into the universe today than yesterday.

But this applies only to space. It is *not* the case that everything in the universe is expanding. Objects that are held together by forces other than gravity, and small regions of the universe that deviate from the larger-scale uniformity (such as galaxies) do not expand.

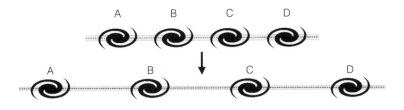

Figure 3.11. Galaxies on a piece of elastic, which is then stretched. Every galaxy moves away from every other galaxy.

We need to think a bit harder about expanding space. Consider a string of galaxies, all evenly spaced. We'll imagine that they are all attached to a long piece of elastic (Figure 3.11). Now stretch the elastic, so that all the galaxies get a bit further apart. Let's sit ourselves comfortably on Galaxy B – what do we see?

After a certain time, the distance from our Galaxy B to Galaxy C stretches by 10%. If it was 10 million light years away before, then it is 11 million light years away now. Similarly, Galaxy A is now 10% further away. Now look at Galaxy D: because all distances are stretched, the distance from Galaxy B to Galaxy D also increases by 10%, from 20 million light years to 22 million light years in the same amount of time. So, relative to us, Galaxy D is moving away twice as fast as Galaxy C.

Because of this motion, light from the other galaxies will be redshifted when it arrives at us on Galaxy B. We can think of this shift in two complementary ways. We can think of it as a Doppler shift, due to the speed of the galaxy as it moves away from us, carried along by the expansion of space. Alternatively, we can think of the expansion of space as stretching the light as it travels from the galaxy to us. According to Einstein's theory of gravity, which also tells us how light moves through spacetime, these two descriptions are equivalent.[8]

The result is Hubble's Law: the redshift of a distant galaxy is proportional to its distance. This is the most important and most direct evidence for the big bang theory. We can see the universe expanding away from us.

But we are not at the centre. Consider our neighbour in Figure 3.11, Galaxy C. They see us in Galaxy B moving away on one side, and Galaxy D moving away on the other side. And Galaxy A, twice as far away as Galaxy B, is moving away twice as fast.

Now imagine a line of galaxies stretching in both directions, with no end. Everyone will see the same thing: every other galaxy is moving away, and galaxies twice as far away are moving twice as fast.

The eagle-eyed reader might have noticed that the relationship between redshift and distance in Figure 3.8 isn't quite a straight line. It starts straight but bends downwards a bit at large distances. And by "large", we mean most of the way across the observable universe. The reason for this is that the expansion rate of the universe isn't constant; it can speed up or slow down. Because light from the distant universe takes time to reach us, we are seeing distant galaxies back when the universe was expanding at a different rate. These galaxies have a different velocity to what we would expect from our immediate cosmic neighbourhood, and thus depart from the linear Hubble Law.

This curve in Hubble's Law can be understood using Einstein's theory, and tells us about the expansion history of the universe. In fact, it tells us something surprising – for the last few billion years, the expansion of the universe has been accelerating. The universe is not just getting bigger; it's getting bigger at a faster and faster rate. We'll return to this in Chapter 10.

A Rival Falls: A Static Universe

At the end of the last chapter, we saw that Olbers's paradox rules out a universe that has *all* of the following characteristics: infinitely large, filled everywhere with stars, infinitely old (so that we can see stars that are infinitely far away), and with distant stars that look the same as local stars. We must reject at least one of these.

Hubble's Law, quite independently of its interpretation in the big bang theory, leads us to reject the final assumption. Light from the distant universe is redshifted. If we look far enough away, we will not see stars at all because their light has been shifted to long wavelengths, too long for our eyes to see.

The observations that establish Hubble's Law are very bad news for the static universe. In such a universe, without any further tweaks, we would not expect to see any systematic trend between distance and redshift.

Can we salvage a static universe somehow? We would need to postulate some other source of cosmic redshift. We have already seen that it would take a magical form of matter to produce the uniform shifting of light we see from distant galaxies. To be able to shift radio waves, microwaves, infrared, visible and ultraviolet light, and x-rays with equal dexterity requires a form of matter that we haven't even seen in our theories, let alone the lab, let alone the entire universe.

We have also seen that the galaxies won't gravitationally redshift their own light – they aren't dense enough. So, to explain redshifts without the Doppler shift (or equivalently, the expansion of space), we need another way of shifting light. Given Hubble's Law, this needs to happen along the way, so that the further the light travels, the more it is redshifted.

This has led to the postulation of so-called *tired light* theories, which postulate that light loses energy as it travels through the universe. This would explain Hubble's Law, but is totally ad hoc. We would like to know how and why light loses energy, but details are scarce. Even so, we will consider such theories in more detail in the next chapter.

So, barring the discovery of an as-yet unknown way of shifting the wavelength of light, the static universe falls. The universe doesn't just sit there. It's expanding.

Anomalous Redshifts: A Problem for Hubble?

Galaxies aren't the only astronomical objects with large redshifts. When the first telescopes that could detect radio waves were built, they discovered very bright, very compact sources in the sky. This is odd, because if you see a single point of light in the sky, it's usually a star. But stars usually don't emit a lot of radio waves. These mysterious sources of light were called *quasi-stellar objects*, or *quasars* for short.

When astronomers were able to identify the source of these radio waves and observe the object using optical telescopes, they discovered that they have exceptionally large redshifts. In fact, for most of the past few decades, the most redshifted objects known in the universe have been quasars. We have since found around half a million quasars, with a wide range of redshifts.

If the light from quasars is redshifted because of the expansion of space, then quasars are a long way away. Therefore, to be seen by our telescopes, they must be very bright. Our best model of how quasars work involves a very hot disk of matter swirling around a black hole, releasing energy as it spirals inward. We'll learn more about quasars in Chapter 6.

(On 10 April 2019 – yesterday, when this was written! – the Event Horizon Telescope team announced that they had taken an image of the accretion disk around a black hole in the galaxy M87. This image is beautifully consistent with this model, showing a hot glowing disk with a dark centre.[9])

But do these objects really fit into an expanding universe? Given that they are very extreme, very weird objects, are we sure that their redshifts are purely the result of the expansion of space? The late Halton Arp, an American astronomer, compiled a catalogue of quasars that appear to be associated with low-redshift galaxies. For example, as shown in Figure 3.12, Arp and his collaborators announced in 2005 the discovery of a quasar with a high redshift (2.114) that appears to be in the disk of a galaxy with a low redshift (0.0224).[10] Obviously, if the quasar is in the galaxy, then it can't be at the distance we would infer from Hubble's Law.

But there's a complication. All we can say is that the quasar *appears* to be in the galaxy. The quasar could be, by coincidence, behind the galaxy and shining *through* its disk of stars, rather than being inside it. How likely is that? What kind of thing is a quasar, if it's not as far away as we thought? What are the consequences for the big bang theory if the redshift of quasars isn't due to the expansion of space?

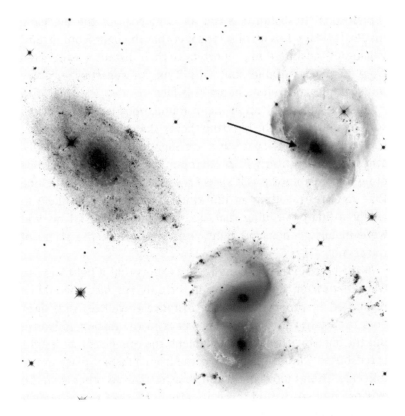

Figure 3.12. Three galaxies of Stephan's quintet, which are located in the nearby universe (with a redshift of 0.0224). The arrow points to a quasar that is apparently in the disk of galaxy NGC 4319. However, the quasar has a measured redshift of 2.114. (Original image credit: NASA, ESA, and the Hubble SM4 ERO Team.)

"The big bang was proved wrong. Again," trumpeted a critic.[11] "If astronomy were a science, this paper would mean the end of the big bang. . . . It's time for set-in-their-way astronomers (of any age) to pack away their big bang assumptions and retire." Why? Because "one of the two major foundations of the big bang is that redshift is proportional to distance. . . . The other major foundation of the big bang is that all redshift is a measure of velocity."

Should the big bang be bothered? Not at all. Let's consider the "two major foundations" cited by the critic. First, that redshift is proportional to distance is not an *assumption* of the big bang model. It's data. It is an observed fact that the light from distant galaxies is redshifted in proportion to their distance, as inferred from Cepheid variable stars and Type 1a supernovae. Some unusual quasars do not change this fact.

And the supposed "other major foundation" is nothing of the sort. It is no part of the big bang theory that *all* redshifts are due to the expansion of space and are therefore a measure of distance. A police officer with a radar gun can measure the redshift of light from your car as it speeds away from him, and conclude that you are travelling at 120 km/h. You are very unlikely to escape a fine by arguing that, according to Hubble's Law, you were in fact a thousand light years away from the scene of the crime.

To be precise, the big bang theory claims that *if* light were to travel to us from a distant place in the universe, *then* it would be redshifted in proportion to the distance travelled. And, light does in fact travel to us from galaxies in the distant universe. But the big bang theory does not claim the converse, that *if* light is redshifted, *then* it must have come from a distant place in the universe. Things move in the universe for reasons other than the expansion of space, and there are other causes of redshift, such as gravitational redshift.

Remember your introductory logic: "if A, then B" is not the same as "if B, then A". For example, "if you get caught speeding, then you will be fined" is not the same as "if you have been fined, you must have been caught speeding". Not all fines are speeding fines.

But don't these anomalous quasars show that Hubble's Law is false? Aren't they exceptions to the rule? No. Let's assume that Arp and his collaborators are correct, and that there really is a local object that has a large redshift. Its redshift must be caused by some effect other than the expansion of the universe. It would have to be an intrinsic redshift, that is, the light was already redshifted when it left the object, rather than being redshifted along the way. In the extreme environment of the quasars, something – call it the *Arp mechanism* – is shifting light.

The crucial question now is: can the Arp mechanism explain the redshift of *galaxies*? The foundation of the big bang theory is not just the redshift of any old astronomical object, but specifically of galaxies because those are the objects to which we can measure distances. When astronomers use the redshift of a quasar to infer that it is a long way away, they are *assuming* Hubble's Law. But when they *establish* Hubble's Law, they use the redshifts of galaxies and distances inferred to stars (and their associated supernovae) in those galaxies. Even if Arp and co. were correct about quasar redshifts, it would only show that quasars don't obey Hubble's Law. It would not show that Hubble's Law is false, because Hubble's Law is established by galaxies. The Arp mechanism would tell us something about the astrophysics of quasars; it wouldn't touch the big bang theory.

Remember why we concluded that the redshift of galaxies is unlikely to be intrinsic, and so cannot be explained by the Arp mechanism, whatever it is. Such large *uniform* shifts of light would require a mechanism like gravitational redshift, which needs extreme, black hole-like conditions. Those conditions are found in the vicinity of quasars – more on that in Chapter 6 – but they aren't found across entire galaxies. We can see galaxies. They're just rotating groups of stars. We can understand the light we receive from galaxies in terms of the combined light from stars. We don't see extreme, intrinsic redshifts from nearby galaxies. There isn't the slightest evidence of the kinds of intense, spacetime-warping conditions we need to produce their redshifts intrinsically.

We are not admitting the reality of Arp and co.'s anomalous redshifts, and we'll return to the question of quasars in Chapter 6 when we discuss the *Lyman alpha forest*. The point here is that anomalous quasar redshifts do not overthrow Hubble's Law. They do not explain why the light from *galaxies* is redshifted in proportion with distance. The big bang does.

4 GOING GENTLY INTO THAT GOOD NIGHT

As we look at galaxies deeper and deeper into the universe, their light is increasingly shifted to the red end of the electromagnetic spectrum. This *redshift* is neatly explained by the expansion of space, described with the mathematics of Einstein's general theory of relativity. But, as we pointed out in the opening chapter of this book, explaining what is observed (and by explaining, we mean meaningfully matching the observed data with your mathematical model) is only the first step. We have to look at *all* of the experimental and observational data. And we have to make predictions for future observations, finding new ways to make our theories face up to the real universe.

In this chapter, we focus on observations of time and distance in the universe. We will look at two sets of data about the universe: how clocks are slowed, and how lights fade.

The Slowing of the Clock

Time is a rather rubbery thing within the theory of relativity. Of course, there are physical clocks, and an observer who looks at their own clock will see it tick by at one second per second. But, according to Einstein, when we compare *two* clocks, at different places or moving at different speeds, they may be running at different rates. As measured by another clock, each tick of the first clock may take more or less than one second. The precise difference between the two ticking clocks depends on their motion and the pull of gravity.

This *time dilation* may sound like science fiction, but it has been experimentally confirmed. Given the precision of modern clocks and computer hardware, time dilation is relevant to our daily lives. As we bounce light signals between the surface of the Earth and artificial satellites, it matters whether the clocks down here stay in sync with the clocks up there. In particular, the Global Positioning System (GPS) would be hopelessly inaccurate without our knowledge of the clock-slowing effects of gravity.

What if we saw a clock in another galaxy, receding in expanding space? Suppose that the light from the galaxy has been redshifted by, say, 25%. Then, if we could see an ordinary ticking clock in the galaxy, we would see it running 25% slower. Its hour hand would appear to take 75 minutes to move from 1 pm to 2 pm. *All* physical processes would appear to run slower in this way – it wouldn't matter if we saw a grandfather clock with a pendulum or a digital clock.

This means that we should be seeing the distant universe in slow motion, and the further away a galaxy is, the more slowly we should see its clocks tick.

Unfortunately, as well as not providing standard rulers, galaxies don't come with huge clock faces. (If galaxies did have huge clock faces, we wouldn't have waited until Chapter 4 to bring it up.) Astronomers need to find another way to tell the time, and for this they call again on giant exploding stars: supernovae.

Tick-Tock, Supernova Clock

In the last chapter, we saw that some explosions in the universe are straight out of a cookie cutter. Type 1a supernovae all explode in a remarkably similar way.

The precise mechanism behind the explosion is still being worked out, but we think that the short story is as follows. A white dwarf – the hot remnant of a burned-out star – is pulling matter from an orbiting companion. At some point, a threshold is crossed: the white dwarf accumulates too much matter and is unable to hold itself up against gravity. It will collapse, reignite, and explode.

Exploding stars are complex: gravitational plunge, runaway nuclear reactions, outpouring light and heat. The dying star brightens immensely as it rips itself apart, and then fades steadily as the debris from the explosion spreads out into space. The explosion is so energetic that new elements are forged, including some that are radioactive.

And yet, observations show that these explosions are all quite similar. In particular, they all brighten and fade over a matter of weeks. That's a ticking clock. Nature has given us a period of time – how long Type 1a supernovae take to brighten and fade – that we can compare for supernovae from our cosmic neighbourhood to the distant universe.

Of course, it's not quite that simple. By examining nearby supernovae, those close enough that we can ignore the universal expansion, astronomers found that the explosions were not completely identical. Subtle differences in the chemical makeup of the stars were affecting the explosions: some were slightly more luminous, some were less, some faded faster, some slower. This just means that we have to work harder, because if we can understand an effect, we can compensate for it. Astronomers found a clear relationship between the brightness of the exploding star and the duration of the explosion: brighter explosions take longer to fade away.

With this calibration of exploding supernovae in hand, astronomers could now use the duration of a supernova explosion to reveal its intrinsic brightness, and so use supernovae as accurate distance measures in the universe. As we discussed in the previous chapter, dedicated campaigns that combined the Hubble Space Telescope with the largest optical facilities on Earth found and monitored supernovae from the very distant universe.

However, recall that we are interested here in studying the slowing of the ticking of the supernova clock. This *cosmological time dilation* is actually a nuisance in the calibration of the brightness of the supernovae. Astronomers infer the intrinsic brightness of a supernova from the time it takes to fade, but how can we tell the difference between supernovae that are intrinsically slower and ones that are slowed by the expansion of space?

Astronomers have used two techniques to tease apart the two effects. Firstly, we have observed enough supernovae in the local universe – for which cosmological time dilation is very small – to measure the relationship between brightness and intrinsic fading time. Secondly, there is a remarkably useful relationship between the spectrum of a supernova (that is, how much light it emits at different wavelengths) and the fading time. By observing the spectrum of the explosion, we can infer the intrinsic fading time.[1]

So, what do the observations tell us?

Observing the Dilation of Time

We have already recounted the remarkable use of supernovae for testing Hubble's Law. But the search for cosmological time dilation almost came as an afterthought. For our purposes, it is actually a cleaner test of the big bang theory, because there isn't any theoretical wiggle room. The stretching of light doesn't depend on how quickly the universe has expanded, or its constitution, or its geometry. It only depends on the amount of expansion. And so there is only one way that the cosmic time dilation can depend on the redshift (z) of the source of the light: $1/(1 + z)$.

Figure 4.1 shows the best current data, from a paper by Blondin and collaborators.[2] The points with vertical error bars show measurements of the time dilation ratio for 35 supernovae. On the left of the plot are shown 22 local supernovae, with small redshifts. These should be clustered around 1 (that is, no time dilation); and, they are. This reassures us that we know how to correct for supernovae of different brightness.

Having pinned down the supernovae with small redshifts, we consider the 13 supernovae on the right of the plot. The expectation of the big bang model is shown by the solid black line, while the dashed horizontal black line shows what we would expect for no time dilation. The big bang prediction goes nicely

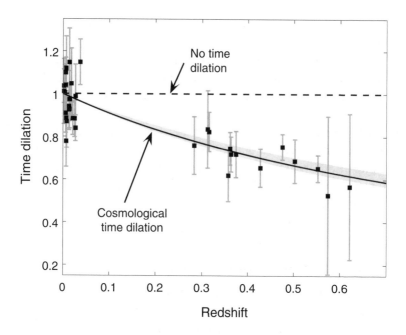

Figure 4.1. Measuring cosmic time dilation. Black squares with error bars are the measurements of cosmic time dilation for supernovae at various redshifts. The shaded grey areas are the lines of best fit. The solid black line shows the prediction of the big bang model, $1/(1 + z)$. The dashed black line shows the prediction if there were no time dilation. The big bang theory fits the data nicely.

through the data, while the *no time dilation* line passes well above essentially all the high-redshift data.

At this point, we return to a central theme in this book: science is more than drawing a line on top of your data and saying "yep, close enough". Scientific data is more than just a measurement; it is a measurement with a degree of uncertainty. The question is whether the theoretical model provides a statistically acceptable fit to the data.

In Figure 4.1, all but one out of the 13 high-redshift data points overlap the theoretical expectation, which is a good sign. But we can do better by using the data to calculate the theoretical curve that *best fits* the data. An appropriate form for the theoretical line is $(1+z)^\alpha$ where α is known as a *free parameter*. The big bang predicts that $\alpha = 1$, and the absence of time dilation would imply $\alpha = 0$. We can try all possible combinations of this parameter and see how well the resultant curve fits the data. The goal is to identify the value of α that best fits to the data, given the uncertainties.

Doing this task with the data presented in Figure 4.1, the result is $\alpha = 0.97 \pm 0.10$. So, we're 68% sure that α is between 0.87 and 1.07 (which is the grey shaded region in the figure), and 99.7% sure that α is between 0.67 and 1.27. This neatly encompasses the big bang prediction but is a long way from $\alpha = 0$.

This kind of calculation is essential if you want to compare competing ideas about how the universe works. We will return to this shortly, when we see how one proposed alternative, the tired light cosmological model, completely fails to match the observed cosmological time dilation. But first, we will consider the tricky question of surface brightness.

The Fading of the Light

Like time, distance is complicated in an expanding universe. We know that more distant objects appear fainter and fainter, but how is this affected by the expansion of space? We are going to explore this in detail by considering the basic properties of light. We need to look at a quantity known as *surface brightness*.

Think back to your science classes; what do you remember about light? You might remember that it travels in straight lines, and that it is an electromagnetic wave, made up of oscillating electric and magnetic fields. You might remember that light is characterized by its wavelength (or corresponding frequency), and that light always travels at the same speed in a vacuum.

One thing you may also know about light is how it dims as a source moves farther away: if you double the distance between you and a light source, its apparent brightness (or *flux*) will drop by a factor of four (two squared). If you triple the distance, the brightness will drop by a factor of nine.

Figure 4.2. A globe with a light at its centre; Australia catches 1.5% of the light. If you cut out the outline of Australia and move it twice as far away from the centre, it only catches a quarter of the light it caught before, i.e. ¼ of 1.5% = 0.375%. The rest, as you can see, streams around the cut-out.

Let's think a bit harder about this law. Imagine a globe, with a frosted light bulb at the centre and all the countries of the world around the outside (Figure 4.2). Light streams away from the bulb and is captured by the globe. The Southern Hemisphere will catch half the light. Australia will catch 1.5% of the light.

Now, take a sharp scalpel and carefully cut around Australia, making a map of Australia and a globe with an Australia-shaped hole. Now, step back, so that you are twice as far away from the light bulb. Light will stream through the Australia-shaped hole in the globe, but because the light is spreading out, the cut-out of Australia can't catch it all. The light has spread out over four times the original area of Australia – twice the height and twice the width. So, the cut-out Australia only catches a quarter of the light, or 1.5% × ¼ = 0.375% of the total light.

Now, think of Australia as being the light-detecting area of your eye, and you can see why doubling the distance between you and the bulb drops its brightness by a factor of four.

But wait ... there's more. Let's remove the globe, and just think about the frosted light bulb. As you step away from the bulb, what happens to its apparent size? How big is the image on your retina? If you remember your geometry, if we have an object with a fixed size, its angular size decreases with distance. If the bulb appears to be 10° across at a certain distance, it will appear to be 5° across from twice as far away.

Because the size decreases linearly with distance, the area will decrease as the square of the distance. If, at a certain distance, the image of the bulb covers 1,000 photo-receptor cells on your retina, stepping twice as far away brings this number down by a factor of four, to 250.

But we already noted that in doubling the distance, the amount of light energy entering your eye every second also drops by a factor of four. Thus, the amount of energy hitting each photo-receptor remains the same.[3] This means that, irrespective of the distance between your eye and the bulb, each photo-receptor in your eye that sees the bulb receives the same amount of energy per second. The photo-receptor registers the same brightness of the patch of the bulb it can see, no matter how far away the bulb is placed. The bulb appears fainter because, as you move away, *fewer* photo-receptors receive light from the bulb.

Putting all this together, astronomers define a quantity called *surface brightness* as follows. Start with a light detector of a given area, which is receiving light from a source with a certain apparent area (that is, the angular area or *solid angle*). How much energy does the detector receive per second, divided by the area of the detector, and by the apparent area of the source? This is the surface brightness.

The important thing about surface brightness is that it is conserved under a wide range of circumstances. For example, if you move away from the source, the amount of energy you receive decreases, but so does the apparent area of the source, so the surface brightness is unchanged. This is not a unique property of your eye, but applies to any light detector, telescope, or camera.

In fact, surface brightness is not changed if you look at the object via a shiny mirror, even if the mirror is distorted. Suppose we take a spherical light bulb to a fun house. Firstly, we hold it in front of a regular, flat mirror, and observe its apparent size and apparent brightness. Now, take a step to the side, and look at the bulb in a mirror that curves towards us in the middle, like seeing a reflection from the outside of a tin can. The bulb will appear thinner, because the area on the mirror that reflects light back to us is smaller than before. But for the same reason, less energy from the bulb will be reflected to us. So, the surface brightness, which depends on the ratio of these two quantities – received energy and apparent area – remains the same.

Lenses have the same property. This can be a tricky thing to get your head around, but it is easy to test. On a cloudless day, take a magnifying glass and look through it at a clear blue patch of sky, far away from the Sun.

DO NOT LOOK AT THE SUN, EITHER WITH OR WITHOUT A MAGNIFYING GLASS! WE KNOW ASTRONOMERS ARE ALWAYS SAYING THIS, BUT SERIOUSLY, THE SUN IS VERY BRIGHT. YOUR EYES ARE VERY SENSITIVE AND WILL BE DAMAGED.

Magnifying glasses, of course, take a small region and make it appear larger. Through the magnifying glass, you are seeing a smaller patch of sky than you would without the magnifying glass. Now compare the brightness of the sky through the magnifying glass to a patch nearby. Is it brighter through the magnifying glass? Or dimmer? You will find that the sky looks the same.

The reason is similar to the case of the mirror. When you look through the magnifying glass, the apparent size of the patch of sky increases, and energy from a larger area of sky is being collected and focused onto your eye. As before, the ratio of the received energy to the apparent area remains the same, and so the surface brightness is unchanged.

Consider another example. Some naughty children have been known to use a magnifying glass to focus the Sun upon helpless ants, frying them to a crisp. An ant does not last long in the focused light of the Sun, but we can consider what they might see in their final moments. Looking up, through the magnifying glass, the ant sees the Sun. The brightness across the disk of the Sun is the same as before, but now the Sun appears to be huge, stretching over most of the sky. To the ant, the magnifying glass has not made the Sun hotter or brighter, but it has brought the Sun an awful lot closer. It's a lot for an ant to think about in its final moments. (Of course, leave the poor ants alone. Watch the movie *Honey, I Shrunk the Kids* if you need to increase your empathy for ants.)

Cosmological Complexity

Having told you that surface brightness is usually conserved, we have to talk about when it's not. Surface brightness is not conserved in an expanding, relativistic universe.

Why? There are several reasons for this, and we will go through them one by one. The first is redshift.

As we have seen, light is redshifted as it travels through our expanding universe. This means that, by the time light enters a telescope or your eye, each particle of light has less energy than when it was emitted. Now, surface brightness is dependent upon the energy entering your eye every second, and, due to cosmological redshift, this energy has dropped. For example, if the light from a distant galaxy has been stretched by 20%, then each photon has lost 20% of its energy. And so we expect the galaxy's surface brightness to be dimmed by 20%. More precisely, if the redshift of a galaxy is represented by the variable z, we divide the surface brightness by $(1 + z)$.

But there's more! As we saw in the previous section, the expansion of the universe also results in a cosmological dilation of time. If our distant galaxy is emitting a certain amount of energy every second, when that light arrives at Earth, this emission is now spread over $(1 + z)$ seconds. We receive the light at a slower rate, and this further dims the surface brightness of a distant galaxy by another factor of $(1 + z)$.

The final effect is a little trickier to explain. Remember that light spreads out as it travels away from a bulb. But in an expanding universe, there is also the expansion of space itself. Light spreads out as it moves *through space*, and, on top of that, space *itself* is expanding. As a result, our detector on Earth misses more of the photons because of expansion. However, when we measure the angular size of the galaxy, we are looking back to the past, to the time when the universe was smaller. The angular size, paradoxically, is not affected by expansion. As a result, the expansion of space further dims the surface brightness of a galaxy. This effect contributes two additional factors of $(1 + z)$, one for the height and one for the width of the area of the detector.

Putting it all together, we expect that, in an expanding universe, surface brightness declines as $(1 + z)^4$. Put another way – suppose we filled the universe with identical galaxies. If the universe were not expanding, then all the galaxies would have the same surface brightness; we would receive less light from more distant galaxies only because they appear *smaller*. But in an expanding universe, a more distant galaxy would have a smaller surface brightness. If we see galactic light that was emitted when the universe had half its current size, the galaxy's surface brightness would be 16 (that is, 2^4) times smaller.

So, what do we actually see?

To the Very Hearts of Galaxies

To perform this *surface brightness test* (also known as the Tolman test, after cosmologist Richard Tolman), it would be convenient if every galaxy in the universe were identical. Then, we could measure the redshift and surface brightness of a whole bunch of galaxies and see whether they show the tell-tale signature of expansion.

But alas! The universe refuses to provide such a simple sample of objects. Galaxies are all different, and so we have a conundrum – is a given galaxy dimmed by expansion or is it just *intrinsically* fainter? Can we find a useful subset of galaxies, so we can perform our test?

Galaxies are generally classified into one of three groups based upon their shape: elliptical, spiral, and irregular. Roughly, elliptical galaxies are featureless balls of stars, spiral galaxies are flattened disks of stars and gas, and irregular is a catch-all term for everything else.

Galaxies also come in a range of sizes, from the immense cD galaxies at the cores of galaxy clusters, often harbouring more than a trillion stars, to the paltry dwarfs with barely a billion stars each. This range of shapes and sizes means that galaxies exhibit a broad range of intrinsic properties, including – alas – surface brightness.

But there is hope! Astronomers studying elliptical galaxies in the local universe discovered that there is a relationship between three observable quantities: their size, their average surface brightness, and the speeds of their central stars. This famous and much studied result is known as the *Fundamental Plane*. If different galaxies in the local universe are nevertheless united by this relation, perhaps we could use them throughout cosmic time to check for the expansion of space.

The study of this kind was presented in 1996 by Pahre, Djorgovski, and de Carvalho.[4] With observations of elliptical galaxies, they used the Fundamental Plane to infer their expected surface brightnesses.

The result of all this effort is shown in Figure 4.3; let's examine each of the pieces. The dots with vertical bars are the raw measurements of the surface brightnesses of galaxies, with their associated uncertainties in measurement. There are four theoretical curves superimposed; note that one data point is needed to calibrate the models, which is why all the lines pass through the data point on the left. It's the other two points that test the models.

- The top dashed line shows the effect in a static universe, with no expansion or redshift.

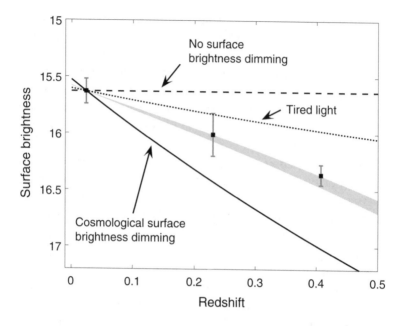

Figure 4.3. The surface brightness test – if we could take the same galaxy to different distances from us (and so, different redshifts for its light), how would its surface brightnesses change? The models are described in the text. Note that the solid "cosmological" line and the dotted "tired light" line fail to pass through the data.

- The solid line shows the expectation for a big bang universe, with surface brightness dimming that changes with the redshift proportional to $(1+z)^4$.
- The grey shaded region shows a range of "Evolution" models considered by Pahre et al. We will discuss the assumptions of this model below.
- The dotted line shows a *tired light* model, in which there is only redshift (no expansion), so surface brightness dimming changes with the redshift proportional to $(1 + z)$. We will discuss this model at the end of the chapter.

Clearly, the big bang – in spite of only (really) having two data points to try to match – does a truly awful job. Not even close. Is this the end of the big bang theory as we know it?

The Problem with Galaxies

As usual, the universe refuses to be neat and tidy for us. The problem with galaxies is that they are made of stars, and stars are not eternal: they are born, live, and die, and each generation is built from some of the remains of previous generations.

The first stars in the universe were made of only hydrogen and helium, because that's all that was available – more on that in Chapter 7. As they ignited nuclear reactions in their cores, lighter elements were forged into heavier elements. As stars age and die, they send some of their gas back into interstellar space via stellar winds and supernovae. This material will be incorporated into the next generation of stars, meaning that their gas is not pristine hydrogen and helium, but is now polluted with the heavier elements.

Successive generations of stars will be more and more chemically enriched, influencing how their cores burn, how their surfaces glow, and how their structure evolves. And so, older galaxies made of older stars will look different to younger galaxies made of younger stars. We wouldn't expect galaxies to look the same at different eras of the universe. Does this make the surface brightness test impossible in practice?

Astronomers are a hardy bunch: if we can understand this effect, then we can compensate for it. Stars are quite well understood; our theoretical models of how stars are structured and how they evolve do a good job of explaining our observations of the heavens. With these models, when we observe a galaxy in the distant universe – which, of course, means that we are seeing it as it was a long time ago – we can ask: if that galaxy was just left alone until today, what would its stars look like?

This is why Pahre and co. focused on elliptical galaxies, which are the retired empty-nesters of galaxies. Unlike their spiralled siblings, they don't have a lot of spare fuel for star formation. Their star-making days are mostly behind them, so their stars are left to grow old gracefully. And that aging process is something we can model quite successfully.

There is another subtlety. The $(1 + z)^4$ prediction assumes that you can measure all of the electromagnetic radiation from a distant source. This would require a device that can collect radio waves, microwaves, infrared light, optical light, ultraviolet light, x-rays, and gamma-rays. Scientists have a name for such a magical contraption: a *bolometric detector*.

Alas, true bolometric detectors do not exist. Different electromagnetic radiation windows require very different detector designs. You can't take a photograph with the radio in your car, and the airport x-ray machine doesn't pick up TV channels.

Why is this a problem? Imagine we had a series of big round light bulbs, scattered through the universe, all emitting white light. If we took photos of the universe, looking at bulbs near and far, would this demonstrate the expected $(1 + z)^4$ dimming?

Actually, no. The problem is the cosmological redshift. This not only dims the light (which we've already taken into account), it also shifts the light out of the range of our detector. A camera doesn't capture all wavelengths of light, so a sufficiently distant bulb won't just be dimmed – it will be invisible. Its light will reach us as infrared light, microwaves, or radio waves. Our camera won't see it at all.

When we look at galaxies, the wavelength window of the detector is fixed here on Earth. As we observe more and more distant galaxies, we see light that has redshifted by greater and greater amounts. Or, turning it around, we are looking at light that was intrinsically bluer (shorter wavelengths) when it started its journey to Earth. We may be looking at red stars in nearby galaxies and blue stars in distant galaxies. Hence, again, we are faced with comparing apples with oranges.

But once again, we don't just give up! We can stitch together observations from different astronomical instruments to understand how galaxies emit at different wavelengths. And we can again use our understanding of how stars shine as they age to reconstruct the full range of light from galaxies. Putting these ingredients together, we can correct for the effect of light lost to redshift; this is known as the *K-correction*. After much checking, it seems that no-one can quite remember what the "K" in

K-correction actually stands for, but it is likely to be from the German word *konstante*, meaning constant. (Readers will be delighted to learn that the previous sentence was, by some margin, the most boring in the book. It's a thrill-a-minute from here on.)

So, finally getting to the point, can the data of Figure 4.3 be explained in an expanding universe after all?

Pahre, Djorgovski, and de Carvalho propose "Evolution" models, which incorporate the aging of stars – which in turn are tested by local observations of stars – and estimates of the K-correction, into their prediction of the dimming of galaxies. The result is shown as the grey shaded region in Figure 4.3. The width of the region represents the state of uncertainty in 1996 regarding the rate of expansion of the universe and the amount of matter in the universe.

The expanding universe model works, as long as we take into account these extra factors – stellar evolution and K-correction. So, is this legitimate, or are we just fudging the numbers?

Manipulations and Corrections

To some non-scientists, all this talk of "corrections" sounds like manipulation: we could make any data fit our model with a sufficiently clever set of extra assumptions. We're just reinforcing our preconceived ideas and getting the answers we want.

We need to explain why these kinds of complications are inevitable. It doesn't matter whether you're a fan of the scientific mainstream or a budding cosmic revolutionary, you're going to have to deal with complications.

Let's start with the instruments that do the detecting. Inside an optical telescope, light lands on a detector and excites electrons in an electric circuit, which register as an image. We want perfectly sensitive detectors that respond proportionally and predictably to every incoming particle of light, regardless of their wavelength. Alas, the perfect detector doesn't exist. If you want to understand the light that actually came from the sky, then you need to account for – and correct for – the limitations of your instruments.

Remember that your eyes are doing similar corrections every day of your life. Your brain performs real-time image processing to provide you with the information you need about the world.

With astronomy, the challenges are compounded by several factors. Firstly, the things that we are looking at are really far away. That sentence puts an unrealistic burden on the word *really*. It takes a concerted effort to mentally picture a factor of 10,000 in size – for example, you can just about imagine a line of sugar cubes (1 cm) stretched along a football field (100 m), holding them all in mind at once. Call each mental leap of 10,000 a single *step*. Starting with our sugar cube, a football field is one step away. Another step takes us from a football field to 1,000 km, the length of the United Kingdom or the width of Texas. You'll need three more steps to get out amongst our neighbourhood of stars, another step to get to our nearest galaxy, and one more step to reach the edges of the observable universe. Needless to say, we can't look at something out in the universe from a different perspective, to get a three-dimensional picture of what's going on.

Secondly, the light that we see was produced by a wild menagerie of physical processes. Take a simple cloud of hydrogen gas. Now, we have one burning desire: to look at it. That can't be too difficult, right?

Well, it depends on what the hydrogen is doing. A hydrogen atom is just a proton with an electron attached (we call this *atomic hydrogen*). If the hydrogen is cold enough, two hydrogen atoms will join together to make a hydrogen molecule (*molecular hydrogen*). But if the hydrogen is very hot, or bathed in high-energy radiation, then the electron will detach from the proton and both will float freely (*ionized hydrogen*). We illustrate these in Figure 4.4.

If you want to see atomic hydrogen, grab yourself a radio telescope. The electron orbiting in a hydrogen atom will occasionally emit a radio signal with a wavelength of 21 cm. The amount of 21-cm radiation depends on how much atomic hydrogen is in the cloud, but it also depends on the temperature of the cloud. Where is the cloud getting its energy from? How is it

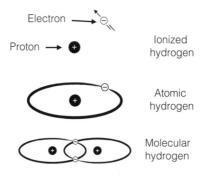

Figure 4.4. Three forms of hydrogen. At the top, a free proton is called *ionized* hydrogen; the electron is unbound. In the middle, when the electron is bound to the proton, we get *atomic* hydrogen. When two hydrogen atoms are bonded together by their electrons combining their orbits, we have *molecular* hydrogen.

distributed through the cloud? You need to know these factors to connect the amount of light to the amount of hydrogen.

What about molecular hydrogen? This is quite difficult to see at all, because – for various technical reasons – the electrons in molecular hydrogen need a very energetic shove in order to be induced to radiate, a shove that they rarely get in the cold environments where they live. We know this because we have studied hydrogen molecules in laboratories on Earth.

Ionized hydrogen is more promising. With plenty of free electrons flying around, bonding and breaking bonds with protons, the gas will glow in visible and ultraviolet radiation. Again, the amount of radiation will depend on the temperature of the gas and any external radiation that the gas is bathed in.

So ... get on with it! How much hydrogen is there? How do we interpret what our telescopes are seeing?

Well, we can infer the total amount of mass from the speed at which the gas is rotating and swishing about, and we can infer this from the shape of the 21-cm emission line. But if we can't see the molecular hydrogen, we will need to infer its presence with a theoretical calculation, based on its rate of formation and destruction in the conditions we measure in the cloud. And various emitted lines from electrons as they recombine with protons will tell you how much ionized hydrogen is around. Putting this all together, you can finally calculate how much neutral hydrogen there is, and how much molecular hydrogen there is, and how much ionized hydrogen there is. Approximately.

The message is this: the universe is a puzzle. You don't just read off the answers. It's big and interconnected, and we live in a part that managed to make something complex like us. We can't be in a simple part of the universe, can we?

To put it another way, raw observations are boring. Astronomy would be rather tedious if all we could ever conclude was that we saw another bright patch in the night sky. What's actually out there?

To answer this, we need to use the laws of nature – motion, heat, light, gravity, atoms, nuclear reactions, quantum mechanics, and everything else in our toolbox – to extract the quantities that we are interested in. The steps along the way will involve assumptions and approximations, and good scientists will make these explicit. Where an assumption is unavoidable, we will look for ways to independently test that assumption. Where this is impossible, we will translate uncertainties-in-our-assumptions into uncertainties-in-our-conclusions.

The key point of this digression is this: not all assumptions are created equal. Some are testable. Some are simple. Some are precise. Some are consistent with *all* of our observations. Some are consistent with what we've learned in the lab. And some aren't. If you think that the current science is flawed, then don't just complain about it. Replace it. Do better. Contribute to the literature, explaining your view of the universe and why it is superior. This is how science moves forward.

We stress again that the vast majority of astronomical and scientific literature, more than a century and a half of it, is online and freely available. You can delve into observations of stars and the models of quasars and more; we encourage you to do so. But note that this takes time, patience, skill, and, more than anything, effort. If you don't want to *at least* understand what has come before, scientists are unlikely to listen to you.

Returning to the case at hand, what about these specific assumptions, the ones about the aging of stars, K-corrections, and other mischievous-sounding quantities?

Well, let's be honest. Cosmologists aren't exactly enthusiastic about the surface brightness (Tolman) test. The assumptions that we need to make – in order to have a model that explains the data – aren't unreasonable. Stars age, of course, and we observe the redshift of light from galaxies, so of course this changes how they look to our instruments. But we've had to assume that the galaxies are sedate – they haven't made too many stars recently – and that our knowledge of galaxy spectra is good enough to make the appropriate K-corrections. It's a bit messy.

One of the greatest astronomers of the twentieth century, Allan Sandage, published five papers in the early 2000s that attempted to perform the surface brightness test, concluding decades of observational work. He comments:

> the Tolman test . . . has not been quite as easily done as I set out . . . in 1974. It has required many developments not yet made at the time. However, the test seems to have been successful. The Tolman prediction is verified. The expansion would seem to be real.[5]

A positive review, but not exactly dripping with enthusiasm. So, we'd prefer to say that the big bang *survives* the surface brightness test. It doesn't fail, because there are reasonable assumptions that we can make about the universe that render the data consistent with our theory. But it doesn't pass with flying colours either.

A Rival Falls: Putting Tired Light to Bed

As scientists searched for possible explanations of the redshift of galactic light, they alighted on a simple idea. Some mechanism – often unspecified – affects light as it travels through the universe, steadily robbing each photon of energy. The further the ray travels, the more energy it loses. This loss of energy causes the light to redshift in proportion to the distance to the source, neatly explaining Hubble's Law. This *tired light* mechanism, whatever it is, has fooled astronomers into thinking that the universe is expanding.

To decide between tired light models and the big bang, we need a different kind of test. Does tired light correctly predict supernova fading and the surface brightness of galaxies?

The first of these is easy, as the tiring of light has nothing to do with time dilation. In a tired light universe, if a supernova intrinsically brightens and darkens over a period of 30 days, it will be observed to take 30 days to brighten and darken through Earthly telescopes, albeit with the light redshifted to longer wavelengths.

We can superimpose this prediction on the data we discussed earlier, showing the relation in Figure 4.1. Even examining by eye, we see that the "No time dilation" line passes over the top of all the data. There is a large difference in the steadily changing aging rate of a supernova with redshift, and the horizontal prediction of the tired light universe.

But is there a way to save the tired light idea? Can we modify the tired light model to explain the observed data? Yes, but it will take some mental gymnastics. Suppose there is a cosmological conspiracy that causes more distant supernova explosions to be different from local ones. Remembering that light travels at finite speed, perhaps supernovae in the past were different, taking longer to explode and fade than in the local universe, spoofing the relation we see in Figure 4.1. Or maybe we are in a special place in the universe, and more distant supernovae take longer to brighten and to fade, again in a particularly precise fashion that mimics the $(1 + z)$ time dilation seen in the big bang model.

Well, maybe. But notice an important difference between these assumptions and the assumptions that we made previously. Stellar aging and K-corrections are to be found in any big bang universe. We know we have to take them into account, messy though that process is.

By contrast, there is no obvious connection between tired light and supernovae that change their spots. We wouldn't expect to have to deal with this complication in a tired light universe.

But there is a further problem. There is no reason to expect these effects to obey any particular relationship to redshift. To account for the observed data on the duration of supernova fading, the actual evolution must be very specific; it must be the form expected from the big bang theory. To make tired light and supernova evolution work, we need to *fine-tune* the evolution, dialling in a specific but unmotivated relationship.

In science, the addition of more assumptions is sometimes justified. However, if the *only* reason for adding these extra assumptions is to account for *one* set of observations, and we have no way of independently testing those assumptions, then they are ad hoc. You're not playing the game right. Like a crafty older sibling, you're bending the rules just as you're about to lose.

Let's now turn to the cosmological surface brightness test. How would surface brightness vary in a tired light cosmology? Remember that in the big bang model, four $(1+z)$ terms are required, accounting for the loss of energy during cosmic expansion, the time dilation of the emitter, and two terms to account for the spreading of the beam of light as the universe expands. But tired light only needs one of the $(1+z)$ terms, redshifting energy from photons as they travel. So, the big bang model requires surface brightness to diminish as $(1+z)^4$, whereas the tired light universe has surface brightness changing as $(1+z)$. How do these predictions stand up?

We have superimposed both cosmological predictions on the data in Figure 4.3. A by-eye analysis shows us that the dimming of surface brightness also presents a significant challenge to the tired light idea.

"Ah, but what about all that star-aging and K-correction jiggery-pokery," you might be saying. These will also affect tired light models. But note that their effect is to move the prediction upwards, to give us *less* dimming than we might expect. This would also be the case in a tired light universe, meaning that its prediction is even worse than is shown in Figure 4.3.

As before, a clever person could probably find just the right set of assumptions to save the model – perhaps a rapid burst of star formation across the universe in the past. But this would probably put the model in conflict with other astronomical observations of galaxies and their stars throughout cosmic time, and in any case would be ad hoc. (If you disagree, then explain your model in a paper and send it to a scientific journal!)

The prospects for such a model seem rather dim. It would need to seriously fine-tune the parameters to explain the data. Tired light cosmology doesn't stand up to scrutiny. It is time to put it to bed as a serious cosmological contender.

5 AN EVER-CHANGING UNIVERSE

Like all historical events, the acceptance of the big bang model was a little messy. But if we had to nominate a single turning point, it was the discovery of the *cosmic microwave background radiation (CMB)* in the 1960s. This glow, expected theoretically and discovered accidentally, is one of the pillars of modern cosmology. Let's start with the observations.

Is the Night Sky Really Dark?

We saw in Chapter 2 that the darkness of the night sky is a significant scientific observation. But is it really *totally* dark between the stars? We know that our eyes are only sensitive to a narrow range of wavelengths of light – between 380 and 740 billionths of a metre. The full Moon's reflected light will illuminate dust in our atmosphere, making the entire sky glow. This is a major annoyance to astronomers. In 1974, the astronomer Paul Wild wrote a review of current scientific knowledge about the Sun, beginning with the following scathing review of our closest star:[1]

> *I have the feeling that to most astronomers the Sun is rather a nuisance. [It] halves the astronomer's observing time from 24 to 12 hours, and then during most of the rest of the time it continues its perversity by illuminating the Moon.*

If our eyes were sensitive to other wavelengths, especially the infrared, the sky would glow due to emission from molecules, including water.

What about the view from space? Once we get above the glow of the atmosphere, is the night sky dark? Not quite. Whatever telescope we use, sufficiently distant sources of radiation will blend together into a smooth, background glow. Think about the Milky Way: while it is composed of billions of individual stars, our eyes see an almost smooth glowing band of light.

Imagine that we had a perfect telescope, able to detect even the faintest emission, and optics so sharp that it could resolve individual points of light, no matter how small. It would see the Milky Way for what it is: myriad individual points of light. Surely, apart from the occasional illuminated gas cloud, the sky would be dark.

But even then ... not quite.

Astronomers have looked deeply into the darkness of the night sky, at every wavelength that we can build an instrument to measure, from radio waves to gamma-rays. Across this wide electromagnetic spectrum, one particular window stands out. In this narrow range of wavelengths, the sky glows brightly, and so smoothly that no attempt to resolve it into individual sources has succeeded.

As with many discoveries, a trail of hints preceded the eventual breakthrough. To properly tell the story that leads to Arno Penzias and Robert Wilson's Nobel Prize-winning discovery in 1964, we begin in the 1940s, only a few decades after modern cosmology was born.

A Canadian, Some Stars, and a Cold Photon Bath

In the 1940s, the Canadian astronomer Andrew McKellar was looking rather close to home, in astronomical terms, at our own galaxy. The Milky Way contains innumerable stars – well, *practically* innumerable, of course, because counting three stars per second, it would take about a thousand years to count them all. Between the stars are immense clouds of gas and dust. These clouds are the raw material from which stars form, and into which stellar winds and supernovae release material.

As these clouds radiate their energy and cool, they can reach temperatures just a few degrees above absolute zero. Within these cold clouds, simple molecules such as hydrogen (H_2), water and carbon monoxide will form. If the right elemental mix is present, we find larger molecules such as ethanol and acetic acid.

In these cold environments, molecules will not be broken apart by the relatively gentle collisions with each other, but they will spin and vibrate. In an *atom*, when an electron transitions from a higher energy level to a lower energy level, a discrete parcel of radiation (a photon) is emitted at a specific wavelength. Similarly, when a *molecule* changes the way that it vibrates, it will emit a photon, typically in the infrared to radio part of the electromagnetic spectrum.

The important point for us is that an atom or molecule that *emits* light at a certain wavelength will also *absorb* light very effectively at the same wavelength. Like a window with a red tint, which allows red light through but blocks blue light, clouds of atoms and molecules can block very specific wavelengths of light, corresponding to the wavelengths that they can emit.

Returning to the 1940s, McKellar was observing our own Milky Way galaxy, and in particular examining the clouds of gas and dust that lie between the stars.[2] The study of this *interstellar medium* was quite new at the time. McKellar was examining unexpected absorption lines in stellar spectra. That is, when he looked at light from certain stars, certain wavelengths of light were missing. For example, light at 387.437 nanometres was missing, relative to the light we receive at other nearby wavelengths.

When an astronomer sees one of these *absorption lines*, they consult a handy table of atoms and molecules to see what they've found. It helps if your observations see lines at many different wavelengths, so that you can double check the identification. McKellar found absorption from molecules such as CH (methylidyne) and CN (cyanogen) in a cloud along the line of sight between the star and us (Figure 5.1).

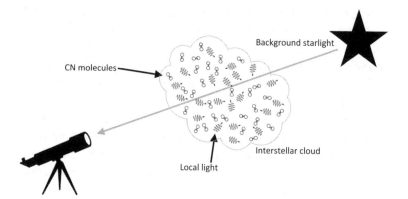

Figure 5.1. Seeing cyanogen (CN) in an interstellar cloud of gas. The local light affects the cyanogen, which in turn affects the light absorbed from the star.

McKellar's important insight came in 1940. Having identified interstellar molecules, he went one step further. Actually, several steps – stay with us.

- *Step 1:* Molecules can spin and vibrate in a number of ways, with different amounts of energy.
- *Step 2:* Molecules swap between these levels when they absorb light.
- *Step 3:* Whether a cloud of molecules absorbs background starlight passing through it depends on what fraction of the molecules are in the right state, ready to receive the background light.
- *Step 4:* The fraction of the molecules in the right state depends on the *local* light, that is, the light *inside* the molecular cloud, which is constantly shuffling molecules around.

Putting it all together, McKellar used his observations of absorption lines to infer the temperature of the light *inside* the molecular cloud. This temperature is very low, but it is not zero. McKellar calculated that it is only about 2.3 kelvin (K), that is, just 2.3 degrees above absolute zero. The molecules were bathed in this light.

It's important to remember that, at this point in scientific history, the big bang theory was just emerging, and nobody connected the presence of this feeble heat with anything cosmological. The observations by McKellar were noted by some, forgotten by others, and unknown to most. A decade later, Nobel Prize-winner Gerhard Herzberg noted McKellar's observations. In his textbook on molecular spectra, he comments that the inferred temperature of the bath of photons "has, of course, only a very restricted meaning."[3] If McKellar's observations had been made a decade or two later, after the theoretical predictions of the hot big bang had become calculated, and if McKellar's observations were broadly known, then he may have won the Nobel Prize for his work.[4]

To summarize, McKellar discovered that there seems to be a ubiquitous, low-temperature bath of photons that pervades all of space, even into isolated, cold molecular clouds. But McKellar didn't provide direct evidence of this radiation – he only calculated its effect on interstellar molecules.

If this radiation is everywhere, could we observe it here on Earth?

The Unavoidable Hiss from Space

In the early 1960s, the American astronomers Arno Penzias and Robert Wilson were trying to use a decommissioned communications receiver in Crawford Hill, New Jersey, as a radio telescope. As with all scientific instruments, they were trying to remove extraneous sources of noise so that it would only receive light from the sky. However, their detector seemed to be picking up a hiss no matter which direction they pointed the telescope.

They thought that this was simply noise in their electronics. But, having battled the electronics, the hiss remained. They then battled pigeons that had made a home in the receiver and cleaned up the copious mess that pigeons make, thinking that this could be the source of the hiss. Still the hiss remained.

Only after they had dealt with all the plausible terrestrial sources of the hiss did they consider the possibility that it was not noise at all. It was a signal. The hiss was, in fact, feeble radiation from space, arriving uniformly from all directions. But they had no idea of this radiation's source.

They phoned a group of cosmologists at nearby Princeton University, led by Robert Dicke. These physicists had worked out that the big bang may have left a radiation aftermath. This radiation would be very uniform but probably have a very low temperature. They were planning to build telescopes to see if this radiation existed.

It must have come as quite a shock, then, when some physicists just 30 miles away rang to say that they had done it already!

To summarize, Penzias and Wilson directly detected the ubiquitous, low-temperature bath of photons that McKellar's observations had suggested. Moreover, they showed that these photons are coming at us from all directions, not from a particular source or even a collection of sources in our galaxy.

However, the communications receiver they used was designed to detect radiation at a single wavelength – roughly 7.5 cm. Cosmologists now wanted to observe this light at other wavelengths. How bright is the sky at other wavelengths?

On the Temperature of Light

In 1994, when cosmologists accurately measured the spectrum of the universe's background hiss –the cosmic microwave background (CMB) – they found a familiar shape. In fact, the equation that perfectly describes the amount of light at different wavelengths had been written down almost 100 years before. How?

We need to review a bit of physics. In particular, what happens when matter (atoms and molecules) and light are mixed in together, bouncing off each other, sharing energy?

At this point, your friendly neighbourhood radio astronomer reaches for a textbook called *Statistical Physics*. Now, if you're not particularly mathematically inclined, adding *statistics* to *physics* might seem like, well, whatever the opposite of "too much of a good thing" is.

But to a physicist, the statistical physics toolkit is extremely useful, and rather elegant. Suppose you want to know how the air in this room will behave if you turn on a heater. The room contains roughly a billion billion billion molecules of oxygen and nitrogen, crashing into each other and the walls and windows. Do we need to predict what every particle will do, solving Newton's equation of motion billions of billions of billions of times?

Thankfully, no. Physicists discovered that they could describe the overall properties of the gas using a relatively small number of quantities. For our room, we just need to know the volume, the temperature, and the pressure, for most practical purposes. If the room is well insulated, and we know how much energy the heater has pumped into the room, then we can work out the change in the temperature.

Why does this work? What exactly are temperature and pressure? In other words, what do we actually measure with a thermometer or a barometer? This is where statistical physics is useful. It tells us that temperature really measures the average speed of each particle. And pressure measures the push that the molecules exert as they collide with the wall.

And we can do more. One of the greats of statistical physics, James Clerk Maxwell, determined the *distribution* of speeds that the gas molecules have. The average speed is set by the temperature; some molecules move faster, some slower.

This variation around the average also applies to the assortment of rotating and vibrating states of the molecules. Let's consider a very simple molecule with only two energy states: a lower energy state, where the molecule is jittering calmly, and a higher energy state, which is jiggling wildly. (This is a major simplification, as real molecules have a great many energy states. But it will do fine for our purposes.)

Now, in a big cloud of molecules, there will also be particles of light. Photons will be absorbed by molecules, turning calm jitters into wild jiggles. And *vice versa*: excited molecules will emit a photon and calm themselves down to the lower energy state. Molecules will be divided between the two energy states.

Just as the molecules have a spread of energies, so will the photons, some at lower energy, some at higher energy. But what mix of energies, exactly? At this point, we need the work of another giant of statistical and quantum physics: Max Planck.

In 1900, Planck was working on this problem, which should be straightforward: matter is made of positive and negative charges, temperature makes things wiggle, wiggling charges make electromagnetic waves (light). Faster wiggles mean more closely packed (shorter wavelength) waves.

But classical physics kept giving the wrong answer. Very wrong. Classical physics said that there should be an *infinite* amount of light, with more and more energy packed into shorter and shorter wavelengths. Something about our classical under-standing of matter and radiation had to change.

Max Planck gave the first clue. "The most essential point of the whole calculation," he said in 1900, was that the energy contained in the matter was, "composed of a very definite number of equal parts." The atoms can't wiggle in any way they please. There is a fixed smallest piece of energy, and each atom has a whole number of them: 1, 2, 3, etc. The light and matter interact as the energy is shuffled and shared between the energy levels.

Using this clue, Planck determined that the energy in a sea of photons was distributed in what is known as a *blackbody spectrum*. The formula he derived does an excellent job of describ-ing the results of experiments. While Planck felt his new deriv-ation was an act of desperation, his insight of *quantizing* the atoms, so that they absorb and emit discrete packets of energy, provided the first steps towards quantum mechanics. Planck started the revolution of twentieth century physics.

Matter and Light, Together

For our purposes, there is another crucial assumption that Planck made in his derivation. He assumed that matter and light were in *perfect equilibrium* with each other. They have the same temperature, and – on average – for every interaction that transfers energy from matter to light, another transfers the same amount of energy from light to matter. In this state of balance, neither is heating or being heated by the other.

Only under these special conditions does the light exhibit Planck's blackbody spectrum. In these circumstances, the light emitted depends *only* on the temperature. This is quite remarkable, given the wide variety of energy levels and molecular transitions and other electron wigglings that produce light. In equilibrium, all this detail is irrelevant, as the matter and light conspire to produce the simple blackbody spectrum.

What if an object isn't in this special circumstance, in perfect equilibrium with the surrounding light? In fact, most objects are like this. A piece of iron at room temperature does not appear to radiate at all, but merely reflects the light that shines on it. But, in fact, an infrared camera would see the iron glowing at longer wavelengths. The light coming from the bar is a mixture of blackbody radiation and reflected light. Almost everything we see is because of reflected light, not its blackbody radiation.

(This, incidentally, explains the term *blackbody*. We imagine an object that reflects no light – it would appear black, even if you shined a torch at it – and ask what radiation it would emit.)

Now place the iron into a hot fire. As the iron heats up, it starts to glow, emitting red light. And as we keep heating, this glow becomes orange then yellow then white as the temperature continues to increase. But even brightly glowing things don't perfectly obey Planck's law. The Sun, for example, is *close* to a blackbody spectrum at 5,500 °C, but not quite. Around the Sun, extremely hot gas at millions of degrees Celsius – called the *solar corona* – isn't in equilibrium with the other outer layers of the Sun, and so contributes more high-energy photons to sunlight than Planck's law prescribes. Other light is absorbed by atoms in the outer parts of the star, atoms that are *not* glowing at 5,500 °C. The Sun isn't a perfect blackbody. Even in a laboratory, it is difficult to manufacture an object that emits a perfect blackbody spectrum.

In fact, the best blackbody spectrum we've ever observed is in space.

On the Temperature of Space

In the decades following Penzias and Wilson's discovery, dedicated experiments measured the CMB emission over a wider range of wavelengths. These experiments used instruments on the ground, lofted in balloons, and launched into space in satellites.

In this book, we want to highlight the *best* measurements of the cosmos, not merely the *first* measurements. So, we will fast-forward to the definitive measurement of the spectrum of the CMB by the Cosmic Background Explorer (COBE) satellite, published in 1994. Onboard that satellite was an instrument with the catchy name "Far-InfraRed Absolute Spectrophotometer" or FIRAS for short. The data are shown in Figure 5.2.

The CMB has a near-perfect blackbody spectrum. It's the best blackbody spectrum we've ever seen. We can't stress this enough: the equation for the grey line in Figure 5.2 was written down 94 years before a high-precision instrument onboard a satellite in space measured the data (black points). The temperature of the CMB is 2.725 K, just a tiny amount above absolute zero. The agreement is astonishing!

But what this result *implies* about our universe is even more remarkable. Remember what Planck had to assume, in order to derive his equation: the light comes from a region where matter and light are in perfect equilibrium with each other, at exactly the same temperature. You can't just point your detector at any old lump of glowing and reflecting and emitting and simmering and wiggling and fluorescing stuff, and expect a blackbody spectrum. And yet, the CMB is not only a perfect blackbody – *it's everywhere*. It's coming at us from all directions, at the same temperature (to five decimal places). There are roughly a billion photons in the CMB for each atom in the universe.

So, we've got the light, but where is the coupled matter, in perfect equilibrium at 2.725 K? Here's the problem: *we know of no*

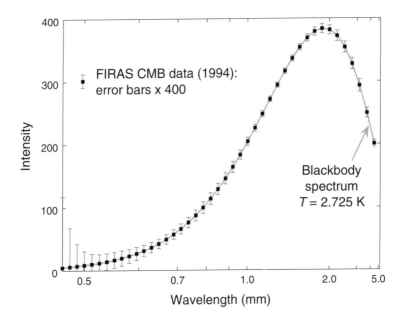

Figure 5.2. The spectrum of the cosmic microwave background, as measured by the FIRAS instrument in 1994 (points with error bars), compared to Planck's theoretical blackbody curve (grey line), which he derived in 1900. With all data, there is an associated uncertainty, which is shown by the vertical error bars. To make the error bars visible on this plot, we have increased their size by a factor of 400!

such matter today. The coldest place in the universe today is the cores of molecular clouds, and these only get down to about 10 K. We can see many of these clouds, but there aren't remotely enough to illuminate the entire universe, even if they all miraculously had exactly the same temperature of 2.725 K.

To make the problem worse, at such low temperatures it would take a long time for matter and radiation to reach equilibrium, even if we could isolate them from the rest of the universe. We need matter and radiation to interact strongly so that they share energy quickly. This would be easier if everything were hotter, because light reacts more strongly with free electrons. And it would be easier if everything were very dense, since there would be more matter to interact with.

Let's summarize: the conundrum is as follows. The universe is filled with radiation. That radiation shows the unmistakable signature of being perfectly coupled with matter at 2.725 K. There is no evidence of any such matter in the universe today.

The Hot Big Bang

In the last two paragraphs, you will find three clues to the solution to this conundrum offered by the big bang theory: (a) there is no perfect matter–radiation coupling at 2.725 K in the universe *today*; (b) it would be easier to create this coupling if the universe were *hotter*; and (c) it would be easier to create this coupling if the universe were *denser*.

The solution: in the *past*, the universe was *hotter* and *denser*. That's the big bang theory in a nutshell. Let's see how it all fits together.

Like all science, the development of the big bang theory was a little complicated, with a mixture of theoretical and observational advances by many scientists. Popular science highlights the big names, in particular Einstein and Hubble, but it's worthwhile examining things in a little more detail.

The key point for this chapter came from the work of the Russian physicist Alexander Friedmann and the Belgian astronomer and catholic cleric Georges Lemaître. If the universe is expanding, they reasoned, then as we wind the cosmological clock backwards, galaxies become closer and closer together. At some point in the past, the distances between galaxies shrinks to zero, and this must have been the starting point for cosmic expansion. The universe was born, in Lemaître's words, from this "primeval atom".[5] (This is a metaphor and shouldn't be taken too literally. The image of a finite-sized ball of matter and energy surrounded by empty space *isn't* a good way to think about the big bang.)

Friedmann and Lemaître's ideas were weird – they still are! – and so the notion of universal expansion and the birth of the universe sat unappreciated by many in physics and astronomy. Some, however, wondered what the early universe would have been like, and wondered if any signatures of the birth of the cosmos remain today. Ralph Alpher, Robert Herman, and eccentric Russian genius George Gamow were thinking about the *thermodynamics* of the early universe, that is, they were thinking about the temperature of the universe in its very earliest of moments. Again, winding the clock of the universe backwards, they realized that because things were packed more closely together in the past, the earlier universe must have been hotter than today. The further back in time we go, the more blueshifted (and energetic) light would be. In the first few hundred thousand years after the big bang, the temperature was high enough to rip electrons off atoms. In the first seconds after the big bang, temperatures were too high for protons and neutrons to stick together and form the nuclei of atoms.

Alpher, Hermann, and Gamow in the 1940s realized that the universe must have been born in an intensely hot state and been expanding and cooling ever since. And as it cools, first protons and neutrons formed, and then atomic nuclei, and eventually normal atoms. The universe's high-energy gamma- and x-rays redshift to lower energies. After 13.8 billion years of expansion, they are still with us today as microwaves, a background faint sea that bathes us all.

Here is a crucial point. At these early times, when the universe was very hot and dense, matter and radiation were in perfect equilibrium. They had the same temperature, and the radiation had a perfect blackbody spectrum. Approximately 380,000 years after the big bang, the universe became cool enough for electrons to bind to nuclei. With fewer free electrons to interact with, matter and radiation ceased to be perfectly coupled at the same temperature. This "decoupling" occurred when matter and radiation had a temperature of about 3,000 K. Since then, the universe has expanded by a factor of over 1,000. But – and here is the crucial point – the radiation kept its blackbody spectrum. The expansion of space simply stretches the photons, which lowers the temperature and leaves us with today's CMB at 2.725 K.

That's why we see *equilibrium* radiation without any matter that it's in equilibrium with. It's a redshifted relic of an earlier time in the history of our universe. The early stages of the universe were the most perfect hot oven that ever existed.

But let's be very clear about one point. The big bang theory *does not* predict the temperature of the CMB. There is no calculation within the big bang theory that produces the number 2.725 K.

The reason for this is simple: in the model, the temperature of the CMB changes with time. Three billion years ago it was 3.42 K, and three billion years from now it will be 2.22 K. The big bang theory doesn't predict the temperature of the CMB *today* because *today* keeps changing.

What the big bang theory predicts, to be precise, is that the radiation that coupled with matter in the early universe would decouple when matter formed neutral atoms, and would exist today as low-temperature, uniform blackbody background radiation. The decoupling would happen at about 3,000 K, and galaxies and stars wouldn't form until the universe had expanded by a significant factor, so a CMB temperature of a few kelvin is in the right ballpark. But that is the best that the big bang theory can do.

Pay attention, cosmic revolutionary. If your theory can neatly and naturally predict uniform blackbody background radiation *with temperature equal to 2.725 K*, then it will have scored a win over the big bang theory.

The reason that the big bang's absence of a prediction for the CMB temperature hasn't counted much against it is because no plausible alternative cosmology has predicted a perfect blackbody spectrum at 2.725 K. Although, some have tried.

A Rival Falls: Steady-State Cosmology

The 1945 film *Dead of Night* is a collection of short horror tales, including a particularly terrifying ventriloquist's dummy that drives Michael Redgrave's character, Maxwell Frere, quite mad.

But one feature of this film led to one of the most acrimonious cosmological battles of the twentieth century.

The story goes that, soon after release, three astronomers, Fred Hoyle, Tommy Gold, and Hermann Bondi, saw the film. They were intrigued by the closing of the film, which brought the stories full circle: the final scenes flowed seamlessly into the opening scenes, so that the movie could be watched on a loop. At the end of the movie, Gold is supposed to have remarked, "What if the universe was like that?"[6]

What if the universe was on a similar loop, in a *steady state*, never changing its average properties over time? Gold was not resuscitating the static universe of Einstein, but rather sought an expanding universe that doesn't change. The idea is that, in a steady-state universe, one epoch of the universe looks pretty much identical to any other epoch of the universe. An observer dropped into the universe at a random place, and, importantly, at a random time, would always be presented with a very similar view: roughly the same number of stars would be shining, with roughly the same colours, with roughly the same number of distant galaxies seen stretching out into deepest space.

They called this the *Perfect Cosmological Principle*: every observer at every time would see the roughly the same properties of the universe. This is an extension of the *Cosmological Principle* that underpins the standard big bang cosmological model, which states that the universe is *homogeneous*, meaning that the universe has the same average properties in every place, and is *isotropic*, meaning that we see the same average view in every direction.

In an expanding universe, the increase in distance between objects constantly dilutes the density of matter and energy, meaning that the universe changes with time. How, then, can an expanding universe obey the Perfect Cosmological Principle, and look roughly the same at all *times*? This presented a significant problem within Einstein's general theory of relativity. But Gold, Hoyle, and Bondi understood the mathematics extremely well, and realized that there was an escape clause in the equations. They introduced a new property to the universe, something they referred to as the *C-field*. The C here stands for creation, and this field did something quite remarkable; it created matter for free. The trio of cosmologists understood that if the rate of matter creation matched the rate that matter was diluted due to cosmic expansion, the average density would remain the same, upholding the Perfect Cosmological Principle. They did not know the details of the mechanism that extracted matter from the C-field, but that could be worked out later. They had a new cosmological model – the Steady-State Model – which implied that the universe was forever expanding, but forever remaining the same.

As we have noted, writing down your cosmological model is one thing; holding it up to nature is another. The steady-state universe passes the observational tests discussed thus far. Galaxies will obey Hubble's Law about redshift and distance. The time of distant events will be dilated, and the light from galaxies will be dimmed.

Nevertheless, such different pictures of the universe must generate different predictions somewhere. The key is that, in a steady-state universe, the universe always looks the same. Because light travels at a finite speed, light from distant sources began its journey to us millions or billions of years ago. Our observations are a time machine that peers into the distant past.

In the big bang theory, the past was very different from the present: more dense and hotter, as we have already seen. Moreover, as we look back in time, galaxies will be less formed, and the first stars will be bursting into light. In this flurry of activity, massive black holes at the hearts of galaxies will feed on gas, gas that swirls and heats and glows intensely before finally being swallowed. In this model, the earlier universe was an active and violent place.

If the universe was in a steady state, however, no matter how far back we looked, the image we see should be like today. The density of galaxies should be about the same, with similar sizes and shapes to galaxies around us in the local universe. And the stars out there should be just the same as the stars right here.

When the steady-state model was first proposed in the late 1940s, the most distant known objects were *only* tens of millions of light years away. Within this local neighbourhood, everything looks pretty much the same. But through the 1950s and 1960s, our view of the night sky expanded rapidly, driven by technological advancement. In particular, radio telescopes came of age and revealed much more than expected. The universe is filled with powerful radio galaxies, many of which blast focused jets of energetic particles into intergalactic space.

Radio astronomers got to work counting the numbers of these radio sources across the sky. They did not have the distances to most of them, as most radio emission lacks the tell-tale signatures of atoms that allow us to measure redshift (as we saw in Chapter 3). However, with some clever statistics, astronomers calculated the expected number of bright and faint sources across the sky in a big bang and in a steady-state universe.

In the 1960s, radio astronomers claimed that their observations showed a universe that evolves over time, with more powerful sources of radio emission existing in the past, and fewer around today. The supporters of a steady-state cosmology were not convinced. The radio observations were uncertain, and difficult to interpret. To them, the radio astronomers had called the race early, without the strong evidence required to consign the steady-state to the cosmological dustbin.[7]

Around the same time, Penzias and Wilson discovered the CMB. This, for many, was a more decisive death knell for the steady-state theory. The hot beginning of the big bang universe could easily account for matter and radiation in perfect equilibrium. But what could create the CMB in a universe that was always and forever just like the universe we see around us today?

Hoyle, steady-state theory's champion, was undaunted. He thought hard about what he needed: something smoothly distributed across the sky, something at a temperature only slightly above absolute, something that had nothing to do with the birth of the universe. And he claimed to have found it. He proposed that the space between the stars is filled with a particular kind of detritus, produced by stars. We know of copious molecular clouds and dust clouds in space, so much so that the view to distant stars is sometimes blocked by their obscuring effects. But this would not do for Hoyle. He needed special space dust, in the form of iron whiskers.

These iron whiskers would be expelled from stars in their old age. As the name suggests, individual pieces must be long and thin for Hoyle's idea to work. Being made of iron, each whisker could line up in the magnetic fields that traverse the Galaxy, like compass needles pointing north. Hoyle reasoned that these iron whiskers would be kept warm at a temperature of a few degrees by absorbing starlight, re-radiating the energy gently in the form of microwaves. To Hoyle, the steady-state theory could account for the CMB as well as the hot big bang theory.

However, most were not convinced. How did stars produce iron in thin whiskers? How could they produce such a perfect blackbody when they must exist in a variety of cosmic environments? Why is their emission so uniform over the sky?

Hoyle was not a man to give up lightly, and he attempted to counter each argument, revising his model as he went. But to most cosmologists, the war was lost, and he was simply applying band-aid solutions to an already dead idea.

Our measurements of the background in more and more detail have revealed tiny temperature variations that beautifully match the predictions of the big bang theory. The iron whiskers of the steady-state universe could not keep up with this deluge of data, resulting in more and more elaborate excuses and ad hoc fixes. Remember that Max Planck said, "science advances one funeral at a time". Fred Hoyle, a giant of twentieth century astronomy, died in 2001. The steady-state universe effectively died with him.

Evolution of the Temperature of the CMB

The CMB puts one more nail in the coffin of the steady-state universe. As we have seen, while the temperature of the CMB today is barely above absolute zero, in the past it was hotter. The big bang theory predicts that the CMB maintains its blackbody spectrum as it cools, and this is governed by a single parameter, the temperature. At each epoch in the universe, this temperature depends on cosmic expansion by a very simple relationship:

$$T(z) = T_0(1 + z)$$

where T_0 is the temperature of the CMB today, and $T(z)$ is the temperature of the CMB at a time in the past when the light travelling to us would be redshifted by an amount z. So, if the light from a distant galaxy is redshifted by 50% ($z = 0.5$), then the temperature of the CMB when the light left that galaxy was 1.5 times higher than it is today (about 4.13 K). But can we actually measure the temperature of the CMB in the past?

Two methods have been explored. The first uses the *Sunyaev–Zeldovich effect*, named after two great Russian astrophysicists. Photons in the CMB can interact with hot electrons, gaining a small amount of energy. Clusters of galaxies are full of hot electrons. If we measure the CMB in the direction of those clusters, the slight enhancement of the temperature of the CMB *here* can tell us what the CMB temperature was *out there* in the cluster.

Secondly, we can use Andrew McKellar's method of observing molecules that sit in the CMB's photon bath. We'll also need a background light to shine through the molecules, in the way that stars shine through the cyanogen in the Milky Way. Thankfully, we have quasars for that. We also need a suitable atom or molecule. Today, cyanogen has just the right structure to be excited by the CMB, which is what McKellar observed. Looking back in time, the CMB may be hotter, so we'll need to find other atoms or molecules that show an observable effect of the photons around them. Astronomers have found that the molecule carbon monoxide (CO) and atomic carbon (C) are up to the task.

Searching quasar light for CO and C is not easy, but the results have been spectacular. In 2011, Pasquier Noterdaeme and his colleagues presented new observations of CO molecules in the

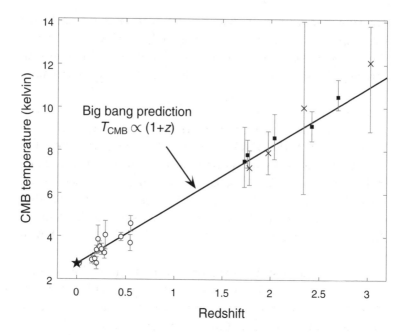

Figure 5.3. Temperature of the CMB in the distant universe, using three different methods. The temperature today is 2.725 K, as we have discussed (the star on the left). The white circles use the interaction of the CMB with hot gas in galaxy clusters. The black squares use measurements of the carbon monoxide molecule. The crosses use measurements of atomic carbon. The agreement with the big bang prediction is excellent.

distant universe, and collected existing measurements of the CMB temperature out to a redshift of $z = 3$; this data is shown in Figure 5.3.[8]

Today, at redshift $z = 0$, the temperature of the CMB is 2.725 K, as mentioned earlier. At low redshifts, the temperature of the CMB is measured from the Sunyaev–Zeldovich effect, while at higher redshift, astronomers call on observations of atoms and molecules. The solid line in Figure 5.3 shows the prediction of the big bang model; specifically, it predicts the slope of the line. Given the temperature of the CMB today – which we measure directly and to exquisite accuracy – the line is an excellent fit to the data.

If you are statistically knowledgeable, you might worry about Figure 5.3. Inevitably, random noise in experiments and observations scatters measurements away from their true values. Remember, error bars, roughly speaking, represent the interval inside which we are 68% sure the true answer lies. So, even a perfect model should only pass through about two-thirds of the error bars. The line in Figure 5.3 seems too good: it passes through too many points, and, in particular, it passes through all nine points on the right. A theoretical prediction that passes through more than two-thirds might indicate *over-fitting*: there is too much freedom, so that your model can twist and turn to fit the variations due to noise, rather than the underlying physical process.

But we have no free parameters! There is nothing to adjust in Figure 5.3. The temperature of the CMB today is measured independently of the other measurements, and the slope of the line is fixed by the model. We cannot shift or bend the line in any way.

Does this suggest that the big bang theory itself is flawed, that the over-fitting freedom is somehow hidden deeper in the equations? A more likely explanation is that the astronomers have overestimated the uncertainties. This is a realistic (and common) possibility, as determining errors is fraught with, well, errors. Scientists would rather be conservative with their error estimation.

The observed relationship between the temperature of the CMB and redshift is a significant success for the big bang theory, and a major blow for the steady-state theory, whose CMB temperature cannot change with time. This supports the idea that the universe had a hot dense beginning and has been expanding and cooling for around 13.8 billion years.

As has been the running theme of this book, if you have an alternative cosmological idea, a model that will overthrow the big bang, then you are going to have to explain not only the existence of the CMB, its uniformity and perfect blackbody spectrum, but also the particular relationship between temperature and redshift.

Good luck!

6 THE WOOD FOR THE TREES

There's a big expanding universe out there, full of astronomical objects: asteroids, planets, stars, white dwarfs, molecular clouds, supernova remnants, planetary nebulae, globular clusters, quasars, galaxies, clusters of galaxies, clusters of clusters. We can see these objects because they shine brightly, lit by nuclear reactions, explosions, and extreme heat. There are also immense regions of darkness and emptiness, cosmic voids that sit between the clusters of clusters.

What about all the matter that isn't so spectacularly illuminated? What can we learn about it? While we cannot see it directly, we can see its silhouette. When we look at a very bright, distant object in the universe, we sometimes see that its light is partially blocked by something between the object and us. So, to see the fainter side of the universe, we require very intense backlights.

And the universe has provided. In fact, we've discovered about half a million of them, so far.

Seeing Radio Waves

One of the most exciting things about turning on a new astronomical instrument, such as a bigger telescope, or one that can observe a new range of wavelengths of light, or one that can survey vast areas of the night sky faster than ever before, is that we don't know what we'll see. Of course, we'll get a better look at galaxies, and discover more stars, and generally see the universe we know in more detail. But hopefully, we'll see some things we've never seen before.

During the Second World War, a large number of electrical engineers developed the skills and equipment to detect and decipher radio waves. Their *radar* technology could turn radio signals into maps of incoming ships and aircraft. After the war, some of these engineers turned their attention upwards. What would we see if we pointed a radar receiver at the sky?

Thus, radio astronomy was born. Just as optical astronomers want a nice dark place to put their telescope, radio astronomers wanted to build their instruments well away from human sources of radio waves, such as AM and FM radio stations. A huge, mostly empty continent in the Southern Hemisphere would do nicely. And so, from the earliest days of radio astronomy, Australia has been a key player.

Radio astronomy is challenging. The mechanism by which a radio telescope detects light is totally different from how an optical detector does it. Your eyes are detectors of optical light, and work via the interaction of a particle of light (a photon) with individual molecules in your eyes. When a molecule in your retina absorbs a photon, its shape changes, which triggers an electrical signal that is carried to the brain. Modern light detectors, such as the one inside your camera phone, work by making electrons jump across a gap in a circuit, triggering a signal.

But your phone's camera can't see radio waves – their wavelength is many millions of times longer than optical waves, and so they pass through your phone's camera without making the electrons jump to trigger a signal. To make a radio telescope, we need to find a material that will be affected by passing radio waves. That turns out to be quite easy: metals contain mobile electrons that can be sloshed back and forth. To make an image of the sky, we need to very cleverly combine signals from a number of metallic antennae.

Now we face another problem: radio observations are intrinsically blurrier. Think of using light waves in a microscope to look at a single bacterial cell. To see the parts of the cell clearly, we need light to reflect off all the different internal surfaces, exploring all its nooks and crannies. Radio waves, which can be centimetres to kilometres long, are too blunt to see any details.

Using them would be as pointless as trying to feel out the cell by poking it with your finger.

There is a similar problem when we look at the sky. If we compare two telescopes of the same size, one that sees optical light and one that sees radio waves, the former will make a much more precise, sharper image.

We can overcome this problem by building larger telescopes. This is much easier for a radio telescope with its large metal antennae or dishes, compared to the precisely sculpted mirror of an optical telescope. And we can be even cleverer, combining signals from a network of telescopes, effectively creating a super-telescope that can see much more detail of the night sky.

Bigger telescopes to detect radio waves are important for another reason. Longer wavelength photons carry less energy than shorter wavelength photons, which means that radio waves, with wavelengths from centimetres to kilometres and beyond, are often very feeble. To date, all of the radio telescopes that have been operating since the end of the Second World War have collected about enough energy to melt a few snowflakes. Radio astronomy is not easy!

Star-Like, Star Bright

So, what did radio astronomers see when they pointed their telescopes at the night sky? Quite a lot! The universe is buzzing with radio waves, and soon astronomers had large catalogues of new, mysterious objects. What were they?

To identify these sources of radio waves, we would like to take a closer look using the sharper image of an optical telescope. But the problem is still the blurry image of the radio telescope. Suppose you're observing a very important radio source called 3C 273 – the 273rd object in the 3rd Cambridge radio source catalogue. We detect a very bright source of radio waves, but the blurriness of the telescope has smeared the signal over a large area. When we look using an optical telescope, there are a large number of objects within the blur. Which of these is producing the radio waves? Which one is 3C 273?

You might think that the definitive answer to this question would await the construction of even larger radio telescopes, which would make a sharper radio image. But, with a bit of ingenuity, radio astronomers managed to measure the position of 3C 273 very accurately using existing telescopes.

Looking up at the night sky, the Moon appears to move across the background stars. Stars and planets are constantly disappearing and re-emerging from behind the Moon. This suggests a clever way of observing blurry objects. When the radio source disappears behind the Moon, its blurry image in our radio telescope will suddenly blink out. By keeping a very close eye on the Moon, we know that the object must have been somewhere on the edge of the Moon when it disappeared. When the object re-emerges from the other side, we'll suddenly see it again with our radio telescope, giving us another set of possible locations, in the shape of a semicircle. The real source must lie on the intersection of these semicircles. If we see the source disappear and reappear a number of times on different nights, we can very accurately pinpoint its position on the sky.

This is a technique in astronomy known as *occultation*. Now is a good time to mention that the word *occult* literally means "hidden". Astronomers rarely, in our experience, dabble in black magic.

In 1962, the stage was set. From the vantage point of the Parkes Radio Telescope in rural New South Wales, Australia, 3C 273 would be occulted behind the Moon four times: 15 May, 5 August, 20 August, and 26 October. Cyril Hazard, Brian Mackey, John Shimmins, and the Parkes Director, John Bolton, used the 64-metre dish to watch the signal from 3C 273 drop from full intensity to zero over the course of about a minute, and then suddenly return to full intensity about an hour later. Meanwhile, Dr W. Nicholson of Her Majesty's Nautical Almanac Office provided very accurate positions of the Moon using an optical telescope.

The result was "the most accurate determination yet made of the position of a radio source," they wrote in a short article in the journal *Nature*.[1] They not only observed the position of 3C 273, but discovered that it is *two* sources: a bright point of light accompanied by a thin stream of material.

These observations are impressive. Sure, the Parkes Radio Telescope is big – it's worth visiting if you ever find yourself 350 km to the west of Sydney. But because of the large radio waves that it detects, its picture of the night sky is still quite blurry. If your vision were this blurry, the Moon would appear to be a single blob of light with no structure or detail at all. Thanks to the Moon's occulting powers, Cyril Hazard and collaborators were able to measure 3C 273's position on the sky to better than one thousandth of a degree. That's like pinpointing an ant from hundreds of metres away.

With the precise position on the sky in hand, Hazard and Bolton wrote to Maarten Schmidt of the California Institute of Technology, telling him where to point the 200-inch (5-metre) telescope at Palomar Observatory. At the position of 3C 273, they saw a bright point of light, which looked a bit like a star. However, when Schmidt measured the spectrum of light of 3C 273 on the 27th and 29th of December 1962, it didn't match any of the "chords of light" emitted by atoms on Earth, especially those atoms we find in stars.

Having puzzled over these observations for over a month, on 5 February 1963 it was suddenly resolved. For no particular reason, Schmidt compared the wavelengths of the light from 3C 273 with some of the familiar lines of hydrogen. They all seemed to be shifted by the same amount, 16%. The source is redshifted! He quickly confirmed that two other lines were the redshifted signatures of magnesium and oxygen, commonly seen in the light from stars and galaxies.

If this redshift is due to the expansion of space, then, by Hubble's Law, 3C 273 is billions of light years away. While it might look somewhat like a star, it is hundreds of thousands of times further away than the stars we see in the night sky. Whatever 3C 273 is, it must be extraordinarily bright. Hazard, Schmidt, and team had discovered *quasi-stellar objects*, a name that was soon shortened to *quasars*.

What Is a Quasar?

As mentioned in Chapter 3, we have now observed over half a million quasars in the night sky. What are they? We'll consider four clues about quasars that we glean from observations.

Clue 1: They're All Redshifted

Every single quasar that we have ever observed has a significant redshift. As with galaxies, these are all *uniform* shifts – every wavelength of every atom we see in quasars has been shifted by the same factor. If the signature of hydrogen is stretched to wavelengths that are double the size, then so are the signatures of helium, carbon, oxygen, neon, magnesium, iron, and every other element and molecule we see.

And they are all redshifts. All half a million of them. We've never seen a quasar that is blueshifted, or whose light hasn't been shifted at all.

Clue 2: They're Like No Star We've Ever Seen

When we look at a quasar with a spectrograph, which measures the different wavelengths of light, there is no mistaking a quasar for an ordinary star. Recall from Chapter 3 that each atomic element has a specific signature of light that it can imprint, either in emission or absorption. We have become very familiar with what stars look like to a spectrograph: we generally see an overall blackbody shape, like the one we saw in Figure 5.2, but for temperatures between about 3,500 K for small stars and 50,000 K for massive stars. This means that most of the light from stars is emitted in the visible and ultraviolet range. As illustrated in Figure 3.3, light from stars shows absorption lines: specific wavelengths of light are missing, indicating the presence of various elements in the outer atmosphere of the Sun, including helium, oxygen, carbon, and neon. We show the spectrum of the Sun in Figure 6.1.

As we saw in Chapter 3, we sometimes see bright emission lines in spectra. These are rare in the spectrum of the Sun, with only a few found in the ultraviolet. This tells us that most of the

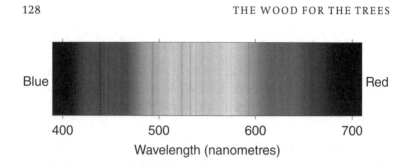

Figure 6.1. The spectrum of the Sun. A smooth rainbow of colours is punctuated with absorption lines (thin grey lines).

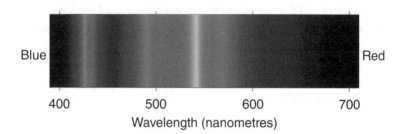

Figure 6.2. The spectrum of a typical quasar. A smooth rainbow of colours underlies bright emission lines.

outer atmosphere of the Sun isn't hot enough to make these elements glow brightly, so they make their presence known by absorbing the light from the star.

We can compare these stars with observations of quasars, shown in Figure 6.2. This is a typical spectrum,[2] observed as if the quasar was at redshift $z = 2.5$. The most important difference is the bright lines where extra light is being emitted. These lines can be many times brighter than the smooth baseline of the spectrum. We occasionally also see dark absorption lines (not shown). Unlike stars, a quasar's emission and absorption lines are usually very broad. We've studied hundreds of thousands of stars with spectrographs, and very few have emission lines like this.[3]

These observations tell us that, whatever a quasar is, it must be very hot, much hotter than stars. To excite atoms into producing broad, bright emission lines, the material in a quasar must be in an environment that is hotter, bathed in more energetic radiation, and generally more extreme than any star.

Clue 3: Something in the Quasar Is Orbiting at Colossal Speeds

We can draw some more information out of the spectra of quasars. Remember that the lines are created when electrons in atoms jump between orbits. The larger the energy gap that the electron jumps, the more energy is released in the form of light. More energetic light has shorter wavelengths – the compressed crests and troughs of the wave pack a tighter, harder punch.

In big atoms, a large number of electrons orbit around the nucleus in a sequence of shells. For example, in an iron atom, 56 electrons circle a central nucleus that contains 56 protons. The innermost electrons are held very tightly, close to the attractive pull of the positively charged nucleus.

In some quasars, we see the signature of these *internal* electrons jumping between energy levels. For example, we can see the *iron K line*, which is emitted when an extraordinarily energetic particle of light penetrates deep inside an iron atom and knocks out the innermost electron. As the remaining electrons shuffle down to fill the newly vacated space, light with a very specific, x-ray wavelength is emitted. This is another chord of light.

We know that there are extreme conditions in the central engine of the quasar, and so the more energetic the light we observe, the deeper we probe. The shape of the iron K line is very distinctive: it can be neatly explained by a spinning disk, with bluer light from the side of the disk that is spinning towards us and red light from the side that is spinning away. (This is in addition to the overall redshift of the light.) But when we calculate the speed at which the disk is spinning, we find that it is close to the speed of light. Something in there, something close to the action, is moving extremely quickly. Which also means that something in there is *holding* the material in a tight, swift orbit.

Clue 4: Quasars Flicker

The night sky is famous for its constancy. The stars and planets shine night after night as they trace the regular paths through the heavens that the ancients admired millennia ago. Even cosmic anomalies such as comets move on a steady path. When we look at a galaxy, the combined light from all the stars is nearly perfectly steady, unchanging from night to night, month to month, and year to year.

Quasars are quite different. Their brightness can vary dramatically over the course of a few hours, and with no apparent rhyme or reason. This is important because it tells us something about the size of the quasar's central engine. Consider a galaxy and suppose that some event caused all its stars to burn a little more brightly. Whatever this event is, its effect cannot travel faster than the speed of light, and so it would take hundreds of thousands of years to reach the other side of the galaxy. The galaxy would brighten very slowly, as each star receives the command to turn up their light source.

So, if a quasar is able to vary its brightness over the course of a few hours, then light must be able to cross the central engine of the quasar in about that time. We can't infer much about what's going on in the centres of quasars from this fact, but we do know that it is happening over a remarkably small region. A typical quasar is as bright as an entire galaxy, and yet all of this light is being produced in a region that is about the size of the Solar System. To get an idea of how small this is: it would be like finding an object that is as bright as the Sun but the size of a car.

The Chief Suspect

What explains all the clues? The most popular model today proposes that quasars are powered by a central black hole. As matter makes its final plunge towards the black hole's *event horizon*, the point of no return, it is accelerated to enormous speeds and crushed in the converging flow. Most of the matter doesn't hit the black hole directly but is swung around

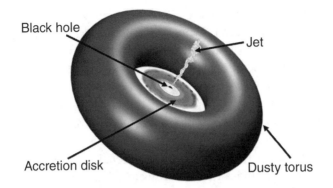

Black hole

Jet

Accretion disk

Dusty torus

Figure 6.3. A schematic view of the central regions of a quasar, with a supermassive black hole surrounded by a high-speed accretion disk and dusty torus.

by gravity into an *accretion disk* of extraordinarily hot and dense matter.

This explains where the light comes from, why it is so intense, and why it is so different from any known star. It also explains why quasars flicker: any interruption to the flow of matter onto the disk and black hole will cause the engine to sputter. The very hot disk explains the iron K lines, and from the rate of flickering we can guess that the disk is a few light hours in size. A schematic diagram of our picture of the quasar's inner engine is shown in Figure 6.3. There will be some redshift of the light from the quasar as it departs the vicinity of the black hole, but not nearly enough to explain the enormous redshifts of most known quasars. So, we infer that quasar redshifts, for the most part, are a consequence of the expansion of space.

Alternative ideas about how quasars work will need to explain all of these observed facts. In the first few decades after their discovery, a variety of quasar models were tried. Perhaps they are *enormous*, super-luminous stars. But stars are quite stable and wouldn't flicker like quasars. Any kind of explosion would be too short lived – quasars flicker on *and* off.

Some astronomers have suggested that quasar redshifts are intrinsic, that is, not due to the expansion of space; we saw this idea in Chapter 3. There are a few problems with this idea. We have discovered quasars whose light is uniformly shifted by a factor of eight. This is an enormous shift – if it were an ordinary Doppler shift, the quasar would be moving at 97% of the speed of light. Some have suggested that quasars are ejected from galaxies at these enormous speeds. But, why are they all red-shifted? Some, at least, should be ejected towards us and appear blueshifted. Unless for some unknown and ad hoc reason, they only emit light backwards – perhaps ejected quasars have a very dusty front that hides their intense emission – this idea will not explain why all known quasars are redshifted.

If, on the other hand, the redshift is due to gravity, then the huge redshifts mean that we are seeing light that is just about to fall into a black hole. The closer to the black hole the light is emitted, the more gravitational redshift. If matter gets too close, then it cannot *orbit* the black hole without being pulled in. For a non-rotating black hole, achieving the enormous redshifts of quasars puts the matter past the point of no stable orbits. Explaining the large redshifts of quasars via gravity would require the implausible scenario that *only* the matter orbiting very close to a swiftly rotating black hole emits light.

For the moment, then, we conclude that quasars really are distant beacons in the universe. Their colossal redshifts are the result of their enormous distance, so they must be incredibly bright. For the purposes of cosmology, the interesting thing about quasars isn't really what they *are*, but what they allow us to *see*.

What Did the Quasar See?

If the big bang picture of the universe is correct, then the light from quasars has travelled an awfully long way across the cosmos to arrive in our telescopes. How did the quasar's light make it all that way across the universe without hitting any-thing along the way?

The answer is that *space* is very well named. To see anything, we need the light from the object to reach us largely

Figure 6.4. Light travelling through a fog. Without the fog, you see the mountain. But as the fog sets in, some of the light from the mountain is deflected away from us, and light from other sources is deflected towards us. The mountain fades, replaced by a white cloud.

undisturbed. Think about a gradually thickening fog over a mountain range. We can see a distant mountain because light travels from the mountain to us. But as the fog thickens, more molecules of water are suspended in the air, and it becomes more likely that a given photon of light that is coming in our direction will be intercepted and scattered. We miss out on that photon, and so our view of the mountain is dimmed. As the fog continues to thicken, eventually no light gets through. Instead, we see the light that is scattered off the water droplets inside the fog. This blurs together light from all directions at once, and so we see nothing in particular, just a white haze (Figure 6.4).

Even air molecules occasionally scatter light. Visible light will typically travel 100 km in air before hitting a particle and scattering. A complete "air fog" would take several hundred kilometres of air; we are prevented from seeing that far by the curvature of Earth.

The distance that light will typically travel in a given region before hitting something and scattering is called the *mean free path*. It depends on two factors. The first is the *size* of the scatterers: it's easier to see through a hundred hovering flies in a given region than a hundred hovering seagulls. The second is the *density* of scatterers: it's easier to see through a hundred hovering flies than a hundred thousand.

For the atmosphere, the size of each scatterer is extremely small – the size of an atom, essentially. But the density of scatterers is enormous. There are over a trillion trillion particles of air in every cubic metre of the atmosphere at sea level. Even 100 km up, at the edge of space, there are still a million trillion particles in a cubic metre.

But even the sparse upper reaches of Earth's atmosphere are positively packed with atoms compared to the universe at large. Suppose you want to take a cubic metre of air and create a sample of the typical density of the universe. You will need to take out all trillion trillion particles ... *except one*. The drastic difference in density means that light will, on average, travel a trillion trillion times further in the universe than in air. So, instead of travelling a hundred kilometres, light could travel 10 thousand billion light years before encountering an atom in the universe.

In fact, most of the matter in the universe today is not in the form of neutral atoms, with electrons orbiting nuclei. Most electrons float free, and are a somewhat larger target for light. But even so, light could travel hundreds of billions of light years between encounters with electrons. The entire observable universe, consisting of the distance that light could travel since the big bang, is only about 50 billion light years. Light can travel from one side of the (observable) universe to the other without a worry.

Well, almost. There are three factors that we have overlooked. The first is that the universe was, on average, denser in the past. So, the further back we look, the denser the fog that the light must pass through. For the most distant quasars, the universe was over a hundred times denser than today.

Secondly, the universe is not uniform like the air in the room in which you are sitting. The cosmos is littered with galaxies, stars, and planets, which are much denser than the rest of the universe. If light wanders through a galaxy, it is more likely to have its journey interrupted.

Finally, we have been talking thus far about visible light. Whether light scatters off a particular atom or electron depends strongly on the wavelength of the light. We are all familiar with this – mobile phone signals can be received indoors, x-rays penetrate beneath our skin, and glass windows can block infrared light.

The relevant scatterers in the universe are hydrogen and helium atoms, and free electrons. Free electrons don't much care about the wavelength of the light. But for hydrogen atoms, certain wavelengths of light are near irresistible.

Making Hydrogen Sing

To understand why hydrogen is so fond of certain wavelengths of light, we need to understand *resonance*. The classic example of resonance involves an opera singer and a wine glass. If the singer can sustain just the right high note, the sound can shatter the glass.

This spectacular effect is all the more striking because sound isn't an efficient way of transferring energy. If you wanted to boil a cup of water using sound, you would need to collect all the energy released by screaming at the top of your lungs for an entire year. So, how does the singer break the glass?

Suppose we could zoom in on the glass, and slow down time so that we see the waves of sound rolling in. A wave breaks on the surface of the glass, pushing it a little. The glass is rigid, and so will push back against the wave. Seen in slow motion, the surface of the glass will ripple like the surface of a pond. Waves will travel in the glass, causing it to oscillate back and forth. Some of the travelling waves cancel each other out, but others will reinforce. If you flick a wine glass with your finger, you will hear the ringing of these reinforcing waves. This is the *natural resonance* of the glass.

Now, watch a series of waves break on the glass. The first hits, and the glass sloshes back and forth. At some point in the sloshing, the next sound wave breaks, providing another shove for the glass. The typically net result will be a random series of pushes and pulls on the surface of the glass.

But suppose we match the sound waves to the natural resonance. The first wave breaks, and the glass sloshes back and forth. At just the time when the glass is about to oscillate back again, the second wave breaks. It pushes the glass at just the right moment to *reinforce* the shove from the first wave.

It's a bit like pushing a child on playground swings. You don't push at random, because sometimes you would be pushing forwards when the swing is moving backwards. Instead, you push when the swing reaches the top of its arc, adding your push to the existing forward motion of the swing.

When an opera singer warbles at a wine glass *at the natural resonance*, each sound wave adds its shove to all the ones before. This amplifies the effect of the waves, until the glass can no longer take the strain.

We can make hydrogen sing in a similar way. The electrons orbiting the central proton in a hydrogen atom have wave-like properties, which means that the atom has a natural resonance. In fact, it has many natural resonances, corresponding to different ways of fitting the electron's waves into an orbit around the proton. If light can hit a resonance in hydrogen, which makes the electron jump between the orbits, then it is much more likely to be scattered by the atom.

The most important example of this phenomenon in hydrogen is called the *Lyman alpha line*, written as Lyα and named after physicist Theodore Lyman. Lyα is the first of a series of such lines he discovered in hydrogen, the others being Lyman beta, Lyman gamma, and so on. If the Earth's atmosphere were hydrogen atoms instead of nitrogen and oxygen (the same number of particles), a typical photon from a light bulb would happily travel about 100 km before being scattered. But if you send a photon with a wavelength of precisely 121.567 nanometres, which is about three to five times smaller than the light from your light bulb, it will barely move for all the enthusiastically resonating hydrogen atoms. It won't travel a thousandth of a millimetre before scattering.

Mapping the Universe in Silhouette

Let's review: quasars are extraordinarily bright and compact sources of light. They are quite mysterious, but we think that their central engine is a black hole being fed by swirling and collapsing gas. Even this extreme environment won't explain their large redshifts, so we believe their light is shifted by the expansion of space. This implies that they are very distant.

Would we expect light from the quasar to make it all the way across the universe without being obstructed by anything? Because of the vast, near-emptiness of space, most of the journey will be uneventful. But, if light of just the right wavelength hits a dense patch of the universe – perhaps in a galaxy – then it is much more likely to be scattered. This will look like an absorption line, a dark patch in the spectrum, as we saw in Figure 3.3.

But there is one more thing we haven't considered. As the light travels from the quasar to us, it is being redshifted. Suppose we have a large, merry band of photons of light, of a wide variety of wavelengths. They are created in the frenzied inner engine of a quasar and begin their journey across the universe towards Earth. As they travel, the universe expands, and so each of our photons stretches.

Perchance, they happen upon a patch of hydrogen atoms. Most of the photons pass harmlessly through, but a few happen to have been shifted into resonance. Those photons that have wavelengths close to 121.567 nanometres see the hydrogen cloud as a dense fog, and soon take a sharp turn and depart from the company. Unperturbed, the remainder continue their journey.

A few million light years later, they meet another patch of hydrogen atoms. But when they arrive, all their wavelengths have been stretched again. So, other members of our group of photons, which passed through the previous patch of hydrogen, have been shifted into resonance. They cannot resist the allure of the 121.567 nanometre Lyman alpha line, and similarly depart the company.

Fast forward to the time when our company arrives at Earth and is detected by a spectrograph. When we look at the different wavelengths of light from the quasar, we will see gaps, dark absorption lines that correspond to the missing members of our company. And here's the exciting part: just as we can connect the redshift of a galaxy to its distance from Earth, we can connect the redshift of these absorption lines to the distance of the patch of hydrogen from Earth. We can make a map of the matter in the universe, even the stuff that doesn't make stars or emit light at all.

Remember that the scattering of light depends on the amount of *neutral* hydrogen, that is, hydrogen with an electron attached. If some hydrogen atoms are bathed in energetic starlight, their electrons won't just excitedly jump between different orbits; they'll jump right out of the atom. The patch of hydrogen will consist of free protons and electrons, not atoms. With no atoms, there is no resonance, and no Lyman alpha line.

So, what would we expect a map of the neutral hydrogen in the universe to look like? We know that there are clouds of hydrogen atoms in our galaxy. These clouds are big enough and dense enough to protect their inner hydrogen atoms from the starlight around them. There are also clouds of hydrogen *between* galaxies, in which most of the hydrogen *today* is not in the form of neutral atoms but is instead ionized (recall Figure 4.4). Looking back into the past of our universe, however, these intergalactic clouds of gas will be denser (because the universe as a whole is denser), and there will have been less time for energetic light from stars to break hydrogen apart (the universe is younger). So, in our map of the hydrogen atoms in the universe, we expect to see dense clouds of gas *inside* galaxies, and more and more clouds *between* galaxies as we look further back.

Figure 6.5 shows what we see. It's a long, thin map, to be read from left to right and top to bottom. We are at the top left, and the further to the right and then down the page you look, the further you are looking into the universe. Each horizontal strip shows a straight line through the universe, showing matter – inside and outside of galaxies – seen in silhouette against

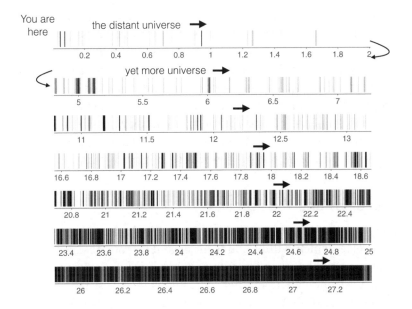

Figure 6.5. A map of hydrogen atoms in the universe, using Lyman alpha absorption. (Take your time with this figure; it's important but a little complicated.) The seven horizontal strips are really one long strip – you should imagine the second strip as attached to the right end of the first strip, and so on, with the bottom strip way off the page on the right. We are located in the top left, and the whole strip is a one-dimensional map through the universe – the distance in billions of light years is written below the strip. As we move further away from us (to the right), we look further back in time, and we see more dense regions of hydrogen atoms (black lines).

a distant quasar. (Actually, we can't get a map like this from a single quasar. Each panel shows a section of map derived from observations of a different quasar.) The numbers under each strip show the distance to the absorbing patch of hydrogen in billions of light years – remember that as we look out into the universe, we are also looking back in time.

What we see is an increasing fog of neutral hydrogen as we look further into the universe. Close to Earth (the top panel), we see a sparse barcode of mostly shallow, grey lines. The quasar shines through small patches of hydrogen atoms, but mostly reaches us untouched. As we look further into the universe, the lines in the barcode become deeper and more bunched. The quasar is shining through a universe that is more closely packed with clouds of hydrogen atoms. The lines start to overlap – they are not perfectly thin – and on the bottom row we see that the light from the quasar is almost completely blocked by Lyman alpha absorption from the neutral hydrogen in the universe.

This sequence of absorption lines, of varying strength and thickness, is known as the *Lyman alpha forest*. We have shown an example above, but the forest is seen in *every* quasar.

Galaxies and the Forest

We can use this map to find the densest hydrogen clouds in the universe, which will be found in galaxies. From observing our cosmic neighbourhood, we know what it would look like if an alien astronomer in another galaxy looked at a distant quasar *through* the Milky Way. There is enough neutral hydrogen in the disk of our galaxy to leave a wide, dark band in the barcode. We expect galaxies to put similar dark bands in the light we see from quasars. We call these hydrogen clouds *damped Lyman alpha absorption* systems, or DLAs. Around 30,000 DLAs have been observed in distant quasars, showing us a population of mostly small to medium sized galaxies.

Because galaxies are where stars live and die, we don't expect them to contain only pristine hydrogen gas; there will be larger elements such as carbon, oxygen, silicon, and magnesium as well. When we see a DLA, we sometimes also see absorption from other elements in another part of the spectrum. In other words, we can detect elements in the gas between the stars in galaxies that obstruct light from distant quasars!

How does our Lyman alpha map of the universe compare with a map of galaxies? This comparison was first done in 2003 by a group of astronomers from Harvard University, the California Institute of Technology, and the University of Cambridge: Kurt Adelberger, Charles Steidel, Alice Shapley, and Max Pettini.[4]

They looked at a set of eight quasars that are a few tens of billions of light years away from Earth, but reasonably close to each other. As well as mapping out the Lyman alpha forest, they were also able to survey the surrounding region for galaxies.

What they found dovetails nicely with the big bang picture. Far away from any galaxy, quasars map out sparse collections of patches of neutral hydrogen. Their spectra show the usual barcode. But as we get closer to galaxies, the barcode thickens. We see more and denser patches of neutral hydrogen.

This is just what we'd expect: galaxies are formed where matter collects, pulled together by gravity. Where you find more matter, you find both galaxies and the kinds of clouds of gas that the Lyman alpha forest maps out.

Also, very close to galaxies, the trend reverses. In the immediate neighbourhood of galaxies, the Lyman alpha forest suddenly disappears again – the barcode becomes more white than black. We can understand this, too: the energetic light from the stars in the galaxy has stripped the hydrogen of its electrons. The hydrogen is there, but it's not neutral hydrogen and so we don't see the resonant Lyman alpha absorption. The Lyman alpha forest map and the galaxy map fit neatly together.

This is what we see in synthetic views of the universe. Using computers, we can create a simulation of the universe at the time of the cosmic microwave background, 380,000 years after the beginning. This is a list of "particles" that represent a certain amount of matter in the universe. Each particle has a mass, a position, and a velocity. We set the simulated universe expanding, as our universe is expanding, and we tell the computer to calculate the force of gravity on every particle, due to every other particle. We then adjust their positions and velocities. Rinse and repeat, and we watch the history of the universe play out on our computer screen.

We include other physical effects, too: thermal pressure from atoms pushing on each other, the cooling of gas as it emits light, the formation of stars in the densest parts of the universe, and the heating of gas by stars, supernovae, and quasars.

Figure 6.6 shows a snapshot of such a simulation. We see a hierarchy of structure of different sizes, known to cosmologists as the *cosmic web*. If we start out with a blob of matter, gravity

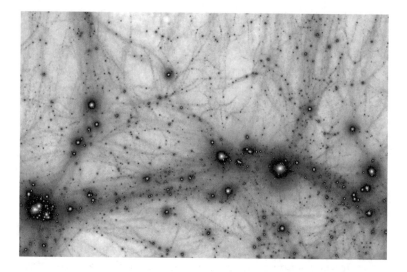

Figure 6.6. The cosmic web in a synthetic universe. As gravity causes over-dense regions of the universe to attract more mass and become more massive still, matter forms a cosmic web of sheets, filaments, and knots. Inside the knots (dark matter haloes), ordinary matter collapses into galaxies. (Simulation by Pascal Elahi, Kevin Lam, and Luke.)

will soon flatten it into a sheet, and then collapse that sheet into a filament. At the intersection of these filaments are dense knots of matter, in which galaxies can form.

We can observe our simulated universe too, calculating what the Lyman alpha forest would look like through the particles in our computer. The result is remarkably consistent with the actual universe – we see a barcode of lines that thicken as we look further out into the universe, further back in time. The assortment of trees in the forest, their bunching and thickness, can be neatly modelled in the context of an expanding cosmic web.

A Contender Falls: Anomalous Quasar Redshifts

Let's review what we've learned, boiling it down to the observational facts. There are very bright points of light in the sky that we call quasars. They are all redshifted, every last one of them, sometimes by enormous factors. Their light is like no star we've ever seen, and something in their central engine is orbiting at colossal speeds, almost as fast as light. Also, they flicker – these intense sources of light can brighten and dim over a couple of hours.

The light we receive from quasars shows narrow absorption lines, corresponding to redshifted Lyman alpha lines – that is, light that resonates strongly with hydrogen atoms. These lines are more abundant as we look at more distant regions of the universe, filling in the spaces in the barcode. Where we see evidence of a galaxy-sized cloud of hydrogen in the forest, we see the expected signatures of other elements as well: carbon, nitrogen, oxygen, magnesium, and more that are produced in stars and distributed into the galaxy by supernovae. And where the forest is thickest, we see more galaxies.

Cosmic revolutionary: your task is to explain all these observational facts.

But note that the precise mechanism that powers the central engine of quasars is not part of the big bang theory. Rewriting that part of the story doesn't change the evidence for the expansion of space and the hot beginning of our universe.

In Chapter 3, we saw that Halton Arp and his colleagues objected to the big bang theory on the basis of anomalous redshifts of quasars. A handful of quasars *appear* to be associated with galaxies at much smaller redshifts. We saw that, however puzzling, this does nothing to change the evidence for Hubble's Law from *galaxies*.

But a closer look at quasars makes it very difficult to imagine how an alternative mechanism (we called it the *Arp mechanism*) could explain their large redshifts. Quasars show uniform shifts – every wavelength, from x-ray to radio, is shifted by the same factor. As explained in Chapter 3, no known interaction of light with matter can cause a uniform shift in wavelength, which shortens our suspect list considerably. Perhaps the Arp mechanism is a pure Doppler shift. But if quasars are local objects in the universe that are moving at close to the speed of light, then we should see quasars moving both towards (blue-shifted) and away (redshifted) from us. But we only see redshifts.

Perhaps the Arp mechanism is gravitational redshift, caused by light escaping from very close to a black hole. But the amount of redshift would depend on how close to the black hole the matter was when it emitted the light we see. We wouldn't see a single redshift but a range of redshifts, as matter falling into the black hole radiated continuously on its plunge. As we saw above, we do see the signature of plunging matter near the black hole and its associated gravitational redshift in the very energetic iron K lines, and these aren't nearly enough to explain the total, overall redshift of the quasar.

We can now add the evidence of the Lyman alpha forest. The Arp mechanism must not only redshift all the quasar's intrinsic emission lines, but must also add a dense forest of thin absorption lines at redshifted Lyman alpha wavelengths. Is there an intricate collection of patches of neutral hydrogen surrounding the quasar? Why are they redshifted relative to the quasar? Given the quasar's substantial output of high-energy radiation, why haven't the atoms in these clouds been stripped of their electrons?

And why are there more and more clouds for higher redshift quasars? This is especially odd if we think that quasar redshifts are due to gravitational effects very close to the central engine. Higher redshifts would indicate that we are seeing light emitted from near the very centre of the action, where even the most tightly bound electrons in a massive iron atom can be knocked clear of the nucleus by the intense radiation. In this environment, what chance does a single electron orbiting a hydrogen nucleus have? We should see less evidence of neutral hydrogen, not more.

And finally, suppose that quasars are local objects, and that the Lyman alpha forest is also some local effect imprinted on the quasar's light. Why, then, would the forest get thicker where its redshift lines up with a distant galaxy? On the big bang theory, the patches of neutral hydrogen in the forest and the galaxy are physically close to each other in space. But if quasars are local objects, then the similar redshifts are mere coincidence. The quasar in our cosmic neighbourhood somehow knows about the redshift of a galaxy billions of light years behind it, from our perspective, and adds a few extra absorption lines.

While quasars are quite mysterious, their redshifts and the redshifts of the Lyman alpha forest fit neatly into our picture of an expanding universe. Yes, there are a few puzzling cases. But *every* quasar shows a Lyman alpha forest that thickens as redshift increases. The as-yet undefined Arp mechanism has millions of quasars to explain, and yet struggles to account for the exceptions, let alone the rule.

As we keep stressing, to be a cosmic revolutionary, you need to explain *all* the data. You can't just focus on the bits that the big bang scratches its head about. There will always be anomalies and mysteries. But to explain five mysteries at the cost of creating a million more, to explain the exceptions but not the rule, is not progress.

7 WE ARE (MOSTLY) MADE OF STARS

The physicist Richard Feynman, when asked what one sentence he would leave for any people rebuilding science after some future apocalypse, replied, "Everything is made of atoms."[1]

Atoms, the building blocks of all ordinary matter, are everywhere in the universe: in stars, in planets, in the pages of this book (or e-reader, or whatever future generations may use). We find 92 different types of atoms (elements) in nature, from hydrogen to uranium, plus a few tens of very heavy elements that scientists have constructed in the laboratory. These basic bits-and-pieces make up all matter, often bonded together into molecules, some with small numbers of atoms, and some with immense numbers. Both molecular heroes – oxygen, caffeine, and alcohol – and molecular villains – sulphuric acid, cholesterol, and hydrogen sulphide – are built from the same finite set of elements.

But where did they come from, these different atoms that make up the universe, including you and me? That is the subject of this chapter.

A Universe of Atoms

The idea that the world is composed out of indivisible pieces has been with us for several millennia, with the ancient Greeks giving us the word atom from the Greek *atomos*, meaning uncuttable. But the theory that matter is composed of atoms was never particularly popular in the ancient, medieval, and even enlightenment worlds. Up until the beginning of the twentieth century, the evidence was ambiguous and opinion was divided.

In fact, it was not long after physicists and chemists put forward convincing evidence of the reality of atoms that we discovered that the atom of chemistry – that is, an atom of one of the elements of the periodic table – is in fact made from smaller pieces still! Atoms are made up of particles called protons, neutrons, and electrons. The protons and neutrons are bound into a dense nucleus, around which the electrons orbit. The difference between one element and another is simply the number of protons in the nucleus. Just to be confusing, in the 1970s we discovered that protons and neutrons are themselves composed of particles called up quarks and down quarks.

A *nuclear* reaction occurs when protons and/or neutrons are added to or removed from a nucleus. A *chemical* reaction occurs when the outer electrons in atoms interact and are shared between them, forming a chemical bond. For example, the energetic reactions of the elements sodium and potassium are due to lone electrons in their outermost orbit, which can easily be shared to make new molecules and compounds.

Chemistry, the study of the elements, their properties and reactions, has a long history. There is no definitive time when alchemy, which sought (among other things) to turn common materials into precious metals, became the modern science of chemistry. Over the centuries, alchemists learned a few things about preparing medicines, purifying natural substances, smelting ores, and making pigments, dyes, salts, glass, and alloys.[2] As the incorrect assumptions of the alchemists were challenged, chemists dissected our material world into various combinations of the 92 natural elements. These were all you needed to construct the material world: animal, mineral, and vegetable.

So much for the Earth. What about the heavens? Are they made of the same stuff?

The Moon looks similar to the Earth in some regards, with mountains, valleys, and plains. The Sun seems like nothing on

Earth, except perhaps fire. Could the Sun be made of the same elements as the Earth? And then there are the stars, distant points of light in the inky black sky. Could we ever find out what they are made of?

In his 1835 book *Cours de Philosophie Positive* (Course of Positive Philosophy), the French philosopher Auguste Comte summarized the state of science in his day. He begins with astronomy, noting the limitations that follow from the large distances to the stars. In particular, since the only human sense that can observe a star is sight:

> *All research which is not finally reducible to mere visual observations is therefore necessarily forbidden to us in regard to the stars. ... We conceive the possibility of determining their shapes, their distances, their magnitudes, and their motions; while we can never study by any means their chemical composition.*[3]

But Comte underestimated the information contained in the light from stars. We can't just look at stars with our eyes and work out what they're made of, but there are more clever things that we can do with light.

In Chapter 3, we saw how the light from the Sun, after being shone through a prism, splits into the colours of the rainbow. The smooth transition from red to green to blue is punctuated by sharp bands of darkness. Joseph von Fraunhofer discovered these lines in the light from samples in his laboratory. While he did not know that they are due to electrons jumping between energy levels within atoms, absorbing energy at specific frequencies, he did realize that each element had its own particular sequence of dark bands, a fingerprint that could be used to identify the elements present within a sample.

This method of splitting light revealed the elemental makeup of the Sun, showing it to be predominantly hydrogen and helium with traces of other elements. As we recounted in Chapter 3, helium was discovered in the solar spectrum via a previously unseen line. As astronomers looked in greater detail at light from the Sun and more distant stars, they found spectral signatures of many elements of the periodic table, from hydrogen to lead and beyond.

The heavens appeared to be very Earthly in their composition!

Measuring the Elements ... in Space!

How much of the universe is hydrogen, helium, carbon, oxygen, and so on? What is the elemental makeup of the universe?

The first place to look is our own backyard. We know that the Earth is mostly iron (in its core) and silicon (in its crust). However, Earth is a small, rocky planet, and most of the matter in the universe is not found in planets. Earth's elemental abundances almost certainly do not reflect the universe as a whole.

Looking further afield, in October 1989, the Galileo space probe was launched towards Jupiter. When it arrived, in December 1995, it was able to directly measure the composition of Jupiter's atmosphere. We discovered that it is about 71% hydrogen, with about 28% helium and 1% heavier elements.[4] As far as we can tell, there are no nuclear reactions on Jupiter, so its atmosphere is a reasonable candidate for reflecting the abundance of the universe as a whole. However, our best models of Jupiter suggest that its interior may be slightly different from its atmosphere, with less helium and more heavier elements.

Most of the mass in the Solar System is in the Sun, so that's a useful place to consider. The same methods that we use to *identify* an element by its light can be used – with a bit of cleverness – to determine *how much* of that element is present. We can see only the outer layers of the Sun, but this is what we want – this gas is the least likely to have been processed by the nuclear fires at the core of the Sun. We find that hydrogen makes up 74% of the total by mass. Next comes helium, accounting for 25%. Most of the remaining 1% is made of oxygen, carbon, neon, nitrogen, magnesium, and silicon.[5]

The Sun uses hydrogen as its fuel – more on that soon. As a result, its elemental abundance might reflect the result of its own element production, rather than that of the universe as a whole. Perhaps the Sun was born from 100% pure hydrogen and has so far burned 26% of its supply. Or perhaps it was born with 75% hydrogen and has only burned 1% of its supply. Can we work out what elemental abundance stars had when they began their lives?

Modern telescopes can determine the elemental makeup of other stars in the Milky Way. Actually, seeing helium in stars at a distance is a bit tricky; it's easiest for very hot stars. Observations of these stars reveal that their outer atmospheres are about 75% hydrogen by mass, varying by 1–2% between different stars. Helium makes up about 23%, with 2–3% heavier elements.[6]

Where else can we look? We can see emission and absorption lines in gas between the stars. For example, the Orion Nebula is a region of gas and dust lit up by some very hot stars. Very deep and detailed observations of the spectra of the Orion Nebula, which can identify over 600 separate spectral lines from nine different elements, show that the nebula is about 73% hydrogen by mass, 26% helium and 1% other elements.[7]

But we can do even better. Some regions like the Orion Nebula – known in the trade as *HII regions* – are bright enough to be visible in other galaxies. Galaxies known as *blue compact dwarfs* are small galaxies being lit up by a burst of star formation. We have good observations for over 1500 such galaxies, which show that they have between 72% and 76% hydrogen by mass, with the remainder almost all helium.[8] Moreover, the fewer heavy elements we find, the closer the abundance of the cloud converges to the ratio of 75.5% hydrogen to 24.5% helium to less than 0.1% other.

These different abundances are remarkably consistent, given the potential complications of astrophysical processes. We don't find any 100% hydrogen clouds or stars, or even 85% hydrogen. There seems to be something *universal* about that 75%–25% mix of the lightest two elements.

Heavy Hydrogen in the Heavens

A crucial piece of information is hiding in the small fraction of pristine gas in the universe that isn't ordinary hydrogen and helium. This section will focus on one specific nucleus: deuterium. A nucleus of deuterium consists of a single proton and single neutron, bound together by the strong nuclear force. An ordinary hydrogen nucleus is just a proton, so deuterium has the nickname *heavy hydrogen*. Unlike hydrogen and helium, deuterium is very rare in all cosmic environments. This is not surprising: the single proton and neutron in deuterium are bound together by the strong force, but this bond is relatively weak (as nuclear bonds go). So, deuterium is easy to either break into pieces or fuse into larger elements. The search for its cosmic abundance has required no small amount of determination.

Astronomers have tried to find deuterium locally in a few ways. Deuterium is not found directly in the Sun, but it is found in the atmospheres of Jupiter and Saturn. This is rather indirect, however: can we be sure that the composition of the atmospheres of these planets reflects the composition of the universe as a whole? Similarly, we find traces of deuterium in local interstellar gas in our galaxy. But again – is this gas representative, or has its deuterium been affected by stars and other energetic astrophysical events? We can also measure the amount of material in the solar wind that is presumed to be the product of deuterium burning, which allows us to infer the amount of deuterium that was around before the Sun was born. This method is an example of what scientists call *theory dependence* – it only works if we are convinced that our understanding of the Sun is correct. But in this book we're trying to give you the raw observations, so we'll look for more direct evidence.

Thus, we turn again to quasars and the Lyman alpha absorbers that they reveal. This is our best hope for surveying a large amount of representative gas in the universe, rather than our highly-processed local environment.

Recall from Chapter 6 that neutral hydrogen is very good at absorbing light that resonates with its electron. Deuterium will do the same – all atoms do. When an electron orbits a deuterium nucleus, we might expect it to be oblivious to that extra neutron, making the absorption properties of hydrogen and deuterium identical. But that neutron does bring mass, which slightly influences the way the nucleus and electron swing about each other. The consequence of this is a subtle shift in the spectral lines of deuterium when compared to hydrogen. While hydrogen will strongly absorb light at 121.567 nanometres, deuterium will absorb at 121.534 nanometres. This 0.03% shift is enough for astronomers to distinguish the two types of nuclei in gas clouds.

So, you might hope to simply point your telescope at the quasar, take a picture, and be done. If only it were so simple: this measurement is difficult. Since the discovery of the Lyman alpha forest, astronomers have searched for the deuterium signature, sometimes without success, and always at the limits of their data.

How do astronomers deal with this kind of uncertainty? Much of science is a battle between signal and noise, maximizing the former and minimizing the latter. Observing for more time and with a larger telescope collects more light from the quasar, boosting the signal. But where does the noise come from?

The first source is the randomness inherent in the counting of photons of light. Even with a steady light source, the actual number of detected photons fluctuates with time. If we expect 100 photons per second *on average*, the first second may deliver 104, the second 107, and the third 98. This is called a *Poisson process*.[9] The relative size of these fluctuations decreases as we collect more photons. When taking a photo with your phone in the daytime, the detector is flooded with photons and so you don't have to worry about fluctuations and statistics and such. But when you are observing a distant quasar and then further sorting what little light you receive into groups by wavelength (i.e. taking a spectrum), understanding your photon statistics becomes crucial to convincing other scientists that you have detected faint absorption lines and thus rare elements.

The other source of noise comes from the sky itself. The atmosphere glows with emission from molecules such as water, and with reflected light from the Moon, or from our own cities, which spend immense amounts of money pointlessly blasting light out into space. (For astronomers, "streetlights that shine upwards" are diabolical villains, whose knavery is matched only by such fiends as clouds, daytime, and people who think we care that they're a Capricorn.) Observations of distant quasars from the ground cannot avoid also collecting light from the atmosphere. While we can subtract the *average* glow from the night sky, the fluctuations of sky are unpredictable and must be taken into account.

And, as ever, there is more! Even after light has entered our telescope, we aren't safe from noise. Sensitive electronic light detectors produce their own noise. Typically, at the heart of an optical or infrared telescope is something called a *charged coupled device*, or CCD for short. Astronomers were among the first to see their potential, because CCDs can be made considerably more sensitive to low light than photographic film. If you're in a laboratory, this doesn't matter much, because you can always turn up the lights. But astronomers can't turn up the lights of the universe, so we need to make the most of every photon that comes our way. For this reason, astronomers have remained at the forefront of research into light receivers, always pushing for more sensitivity.

When a particle of light strikes a CCD, the energy rearranges electrons, effectively placing them in a pile for a given pixel. A photographic image is made by counting the number of electrons in each pile to work out how many photons were collected in particular locations.

So, where does the noise come from? The problem is that the arrival of photons is not the only source of energy that can shift around the electrons. Detectors are not perfectly cold (at absolute zero), so electrons are always jittering to some extent. These jitters can shuffle the piles, adding some electrons, taking some away. So, when you count the electrons collected in a CCD, some are due to the arrival of photons from the quasar, some due to photons from the sky, and some are there, or have been lost, due to thermal jitters.

If your burning desire is to learn something new about quasars, all these engineering issues can be frustrating. But get used to it: observational and experimental science is a constant battle between signal and noise! Failing to grasp this point can make science look like a black art, a magical incantation that brings forth scientific papers in a language foreign to the layperson. But scientists speak the language of mathematics and statistics because this is the language of the signal and the noise. Learn that language: pick up a good (Bayesian) statistics textbook.

So, after all that, how much deuterium is out there in the universe? Fields, Molaro, and Sarkar, summarizing the state of the field in 2017 for an international collaboration known as the *Particle Data Group*, tell us that ten good measurements point to the value $(2.569 \pm 0.027) \times 10^{-5}$ deuterium nuclei per hydrogen nucleus.[10] Meanwhile, by adding a new "near-pristine" absorber in 2018 to their previous sample of six measurements, Ryan Cooke, Max Pettini, and Charles Steidel conclude that the deuterium to hydrogen ratio is $(2.527 \pm 0.030) \times 10^{-5}$.[11] This is about one nucleus in 40,000.

So, do these scientists agree on the answer? They're supposed to be measuring the same thing! In fact, this agreement is pretty reasonable. The uncertainty (one-sigma) regions of each measurement overlap, so there is no reason to attribute the difference between the measurements to anything but the usual observational uncertainties we have discussed.

But which number should we take as the answer? We shouldn't take a single number. We should treat the measurement as a region, a fuzzy section of the number line. We should reach any conclusions about the universe with this uncertainty in full view.

Let's summarize what we have learned about the elemental makeup of the universe. Ordinary matter in the universe is dominated by one particular element, the first in the periodic table, hydrogen. The universe is about 75% hydrogen and about 25% helium, which is the next heaviest of the elements. About 1 nucleus in 40,000 ($\pm$500) is deuterium. Everything else, including the elements that are crucial to human life, such as carbon and oxygen, and less crucial ones such as arsenic and uranium, are present in mere trace amounts.

With the raw facts before us, we ask: why does the universe have this peculiar distribution of elements? What universal alchemists are transforming cosmic lead into cosmic gold?

Alchemist to the Stars!

Of course, we've already met the most abundant alchemists in the universe. Stars burn via nuclear reactions at their cores. Perhaps surprisingly, physicists worked this out only relatively recently.

The source of energy in stars was a long-standing problem in astrophysics. An early suspect was gravity: as the immense mass of the star squeezes its gas, its core heats up in the same way a bicycle pump heats up when inflating a tyre. A physics undergraduate should be able to calculate the gravitational energy released by the contraction of a star, and then use this to estimate how long the star could run on this fuel source. (We stress *should* be able, because when Luke attempted this calculation in an undergraduate physics exam, he got the answer wrong by 24 orders of magnitude, expecting the answer of a few million years and horrifyingly, yet repeatedly, getting 300 picoseconds.)

Gravity could power the Sun for at most tens of millions of years. This is a mighty long time, but not long enough for the many hundreds of millions to billions of years that life has been on Earth.

The discovery of immense quantities of energy within the nuclei of atoms provided a more promising long-term power source for the stars. The star is ignited by gravitational energy, which heats the stellar core to millions of degrees. Nuclei are smashed together at immense speed, bringing them close enough for the strong nuclear force to bind them together. Stars build heavier elements out of lighter elements, releasing energy in the process.

While it was realized that nuclear reactions could power stars for billions of years, the precise details – the sequence of reactions by which small elements are built into larger ones – remained to be worked out.

The heroes of this tale are Hans Bethe, Charles Critchfield, Carl von Weizacker, Ed Salpeter, and a particularly famous collaboration in the 1950s between Margaret and Geoffrey Burbidge, Willy Fowler, and Fred Hoyle. Starting with the work of these scientists, modern astrophysics has pieced together the following picture of how stars are born, live, and die.

Most of the atoms in the Milky Way galaxy are not found in stars, but in interstellar space, in the form of a hot diffuse background punctuated by dense cold clouds. Stars are born when part of a dense cloud, because of its own gravitational pull, fragments and begins to collapse. This piece becomes a protostar, whose intense core heat begins to radiate away the energy of collapse.

The most important property of this protostar – and the star that it will form – is its *mass*. Stars are immense pressure cookers, with the squeeze of gravity heating the stellar core and providing the conditions to fire up and contain nuclear reactions. If the protostar is too small – less than about 10% the current mass of our Sun – gravitational squeezing is not sufficient to ignite nuclear reactions, and the result is a failed star, known as a brown dwarf.

But if the protostar is big enough, protons begin to fuse together and a star is born. The more massive the star, the more intense the gravitational pressure in the stellar core, and the fierier the nuclear furnace. Small stars, known as red dwarfs, are slow burners, running on their gentle nuclear reactions for many trillions of years. Massive stars, the giants, can burn their nuclear fuel in a cosmological instant, a few million years of spectacular life before an extraordinary end.

Stars spend most of their lifetime burning hydrogen into helium in their cores. Our Sun is roughly half way through its 10-billion-year supply of hydrogen, but the day will come when this is exhausted. For small stars, this is the end of the line: once the hydrogen is gone, the nuclear fires simply go out and the star radiates away its remaining heat into the darkness.

But in more massive stars, as the hydrogen supply runs low, outward pressure due to radiation diminishes and the gravitational squeezing increases, driving the temperature upwards and igniting a new round of nuclear burning. In the revitalized star, helium is the fuel and heavier elements the product.

In this new regime, changes need to be made. To adjust to the new balance of inward and outward pressures within the star, its outer layers inflate, expanding enormously. Our Sun, for example, is destined to become a red giant star whose outer layers will engulf the orbit of the Earth. (Stay calm – we've got 5 billion years to come up with a plan. We've got more immediate problems to worry about and, frankly, if humans haven't worked out how to get off this planet by then, we deserve what's coming.)

We see evidence of this process in *planetary nebulae*, though – despite the name – they have nothing to do with planets. We see brightly glowing and expanding shells of gas around older stars, containing elements such as oxygen and carbon. Astronomers love planetary nebulae, because they provide such wonderful pictures! A black and white image will not do them justice, so please find the time to type "planetary nebula" into a Google image search.

For the most massive stars, those at least eight times more massive than the Sun, their nuclear evolution is even more dramatic. As with other stars, they start by burning hydrogen within their cores, but this fuel is rapidly exhausted. In the core, helium starts to burn, while around it a shell of hydrogen still burns. Faster than the hydrogen ran out, the helium fuel in the core is depleted and the core is now mostly carbon. The gravitational squeezing continues to fan the nuclear fires, and carbon begins to burn into oxygen and neon. Still, around the core, there are shells of nuclear burning, an outer one of hydrogen and an inner one of helium.

The process continues, with heavier and heavier elements building up in the core. Soon, the star is layered from heavier to lighter elements, with an iron core, and silicon, oxygen, neon, carbon, helium, and hydrogen shells. But the fun stops at iron.

Iron sits at a special place in the periodic table. Smaller elements, when fused together, produce energy. Larger elements such as uranium do the opposite, producing energy if they fall apart into smaller pieces (fission). But iron, being the most stable nucleus, will do neither. Here, in the middle of the period table, are the energetically stable elements. If you want to do anything with an iron nucleus – fusion or fission – you have to put energy in! There's no spare nuclear energy to be expended.

So, once the core of the star is mostly iron, that's that. In a (cosmic) instant, the core of the star goes out. But gravity's grip is as strong as ever. Lacking support, the outer layers of the star crash inwards, smashing into the core and rebounding. This crash drives the core of the star to immense temperatures – even for the core of a star! For an instant, this injection of energy from gravity reignites the nuclear furnace and elements heavier than iron are forged in the inferno, absorbing some of the energy. These conditions last but a moment as the star rips itself apart in a tremendous explosion, a supernova, throwing these heavy elements into space.

All that hullabaloo – a technical term, meaning a particularly riotous kerfuffle, edging towards pandemonium – is the way that our universe makes and distributes the elements, particularly those heavier than iron. It is sobering to think that a star had to die to make the gold in a wedding ring!

How Do We Know How Stars Work?

In the previous section, we gave the standard, textbook account of the career of stars. It's a marvellous story, but because in this book we are attempting to give you a candid peek behind the scenes of science, we will ask: how do we think we know all that?

Of course, we haven't watched any single star go through all of these stages. But we can see stars, white dwarfs, planetary nebulae, and supernovae at different stages of the story, and see whether we can put these snapshots into a logical order. And while we can't see directly into the core of every star in the night sky, we do know the laws that matter obeys, because we can test them in the laboratory and, in the case of gravity, observe their effect on the Solar System. We can ask what arrangement of matter *both* obeys the laws we know and explains our observations of the universe.

The story of stars has been assembled piece by piece. Scientists tend to focus on one phase of a star's life at a time in their research. Some will focus on the process of star birth, on how gas fragments and collapses at the sites of future stellar nurseries. Others focus on the end points of stellar life, from white dwarfs to neutron stars to black holes. Many study stellar populations, examining the particular mix of giants and dwarfs, and their chemical enrichment and evolution.

Of utmost importance to the whole story is the life of an individual star. Astrophysicists build a model of a star that accounts for the pull of gravity, the push of gas pressure, the flows of energy through convection and conduction, the swirls of gas and plasma, all coupled with the nuclear processes at the stellar heart.

The basic equations are reasonably straightforward, and a lot can be learnt from these on the back of an envelope or two. But in all their magnificence, stars are immense, three-dimensional, rotating and pulsing balls of gas, with temperatures of thousands of degrees on their surface and millions of degrees at their core. They burn for millions and billions of years but explode violently in seconds. And while Astronomer Royal Martin Rees tells us that "even the smallest insect, with its intricate structure, is far more complex than either an atom or a star",[12] traditional pencil-and-paper physics can only take us so far.

So, we turn to a more powerful tool with which we can attack the equations. We can make a living, breathing star in a computer. Computers excel at doing lots and lots of simple tasks, which, taken together, can represent the complex processes inside physical systems. The growth of computational physics and astrophysics has mirrored the growth of computing power in general.

But stars are particularly tricky things to build in a computer due to the *dynamical range* of the problem. Think about watching an interactive video of the life of a star. You could zoom in on the core and its intense nuclear reactions, or zoom in on the surface to see the churning plasma eddies, or zoom out to see the overall structure shaped by gravity. You could watch a flare shoot up in real time or speed up the video to watch the 11-year sunspot cycle. You might be tempted to fast-forward through the millions and billions of years of stable hydrogen burning, but be careful that you don't shoot past the thrilling final act: the ignition of helium burning, pulses of heat that shed the outer layers of gas, and especially the onset of heavy element burning in massive stars – the star might run on carbon for 600 years, neon for one year, oxygen for six months and silicon for one day! And you definitely don't want to miss the supernovae, which will ignite in a second.

To truly understand a star, we need to follow all these processes on a wide range of time scales. Handling such calculations requires special computing skills, ensuring that computational approximations don't get out of hand and ruin the entire simulation.

The first simulations of a star used computers that were much less powerful than your phone, and possibly less powerful than your microwave. The problem was reduced to one dimension, considering only how the structure of the star varies as we move away from the centre. While these models provided insight into stellar inner workings, they lacked key physics that astronomers knew was present. In particular, one-dimensional gas can't create a swirling eddy (*convection*) that allows hot gas to rise and cool gas to fall.

With the steady march of processor speeds and memory, astrophysicists made the leap to two-dimensional stars, which included an approximate model for gas convection in the outer layers of stars. Of course, nature isn't two-dimensional either, so, while providing more insights into the inner workings and evolution of stars, these models were still limited.

The jump to truly three-dimensional model stars has come only in the past few decades, with the advent of supercomputers. These are (usually) built by linking individual processors in a high-speed network, allowing them to talk to each other and share information. While synthetic three-dimensional stars can capture the internal flows of gas and radiation in a brief moment of a star's life, including all the physical bells and whistles – nuclear reactions, conduction, convection, rotation, magnetic fields, turbulence – is still very much a work in progress.

But remember that we're modelling a star, not building one. We must make approximations that balance the model's complexity with the available computer power. There is little point simulating a star if the calculation will take a billion years, although astronomers will patiently wait several months for all of their calculations to complete.

What do we see in our models? They reveal how complex a star can be, especially the convective bubbling of gas in their outer parts, and the interaction between the base of the outer convective parts of the star and the inner, radiation-dominated part, a process known as "hot bottom burning". They show that a red dwarf will burn for trillions of years, steadily mixing the material in its core with the gas in the outer parts of the star. In particular, they show the immense complexity of the final stages and ultimate supernova explosion of a giant star. For example, the final blast can be asymmetric, ejecting the dead core of the star into intergalactic space.

Returning to the topic of this chapter, and the reason for this excursus: these models predict how much of each element will be produced in stars, and how much will be expelled into the wider universe in planetary nebulae and supernovae explosions. This picture of how stars of varying masses live and die feeds into our understanding of how populations of stars, typically consisting of many red dwarfs and few giant stars, are able to turn hydrogen into heavier elements.

Can Stars Make the Elements?

Remember the facts: the universe is 75% hydrogen and about 25% helium. Deuterium accounts for one nucleus in every 40,000. The rest – carbon, oxygen, neon, silicon, etc. – are present in mere trace amounts. We have never found a star or a gas cloud – either in a galaxy or in intergalactic space – with more than 75% hydrogen.

So, can stars alone turn pure hydrogen into the mix of elements we see in the universe today? At least two massive hurdles await any such attempt. Firstly, stars struggle to make enough helium. Secondly, stars *always* destroy deuterium.

Regarding helium, we can do a simple back-of-the-envelope calculation: how long would it take for a typical star like the Sun, starting with pure hydrogen, to turn 25% of its mass into helium? The rough answer is: about 25 billion years.

At the moment, only about 10% of the mass of atoms in the universe is in the form of stars. So, if the universe only cycles about 10% of its matter through stars at a given time, it would take about 250 billion years of stellar burning to turn a pure hydrogen universe into a 75% hydrogen – 25% helium universe. That's a problem, because we can use our knowledge of how stars burn to estimate how old they are, and we have never found a star older than about 15 billion years (plus or minus a few billion years).

To put this another way, all the stars in the universe today, burning for as long as the oldest stars in the universe today, would have turned only 1% of the hydrogen in the universe into helium. Where did the rest of the helium come from?

Deuterium is an even bigger problem: stars only destroy it. When the small amount of deuterium in an interstellar cloud is heated and pressurized in a protostar, it ignites the *first* nuclear reaction. Deuterium quickly reacts with a proton and is used up. When the star begins to burn hydrogen, deuterium is briefly produced in the reaction chain, before being used up in further reactions. A typical proton (hydrogen nucleus) in the Sun will wait 10 billion years to undergo a reaction that creates a deuterium nucleus; that deuterium nucleus will wait about one second before being fused into helium-3. Even the seemingly small amount of one deuterium nucleus per 40,000 hydrogen nuclei is far too much for stars to produce.

If not stars, then where? This is the challenge that awaits any attempt to tell the story of our universe.

The Magic Furnace[13]

How does the big bang explain the elemental abundances of our universe? The story of cosmic alchemy comes in two parts. Over most of cosmic history, stars have turned lighter elements into heavier elements. But – and this is the crucial point – stars don't start from scratch, that is, from a pure hydrogen universe. Before any star shone, the whole universe was once a cosmic oven.

At the moment, we don't have the physical theories and associated mathematical tools to unpick the processes underway at the instant of the big bang. Maybe it was a true birth from nothing, or maybe our universe is the child of an older (and wiser?) pre-existing universe. Nevertheless, there is a classical beginning predicted by Einstein's theory, and we can use this as a useful reference point.

Recall that the early big bang universe was hot and dense, and the earlier we look, the hotter and denser the universe is. There is thus a time in the early universe when matter is hotter than anything we've ever observed in the universe today or created in an experiment. For example, in a 27 km tunnel under France and Switzerland is the world's largest particle accelerator – the Large Hadron Collider (LHC) – which smashes fundamental particles together at enormous speeds so that the debris can be examined. When the universe was about 10^{-14} seconds old, *every* particle in the universe was as energetic as a particle in the LHC. So, before this time, the matter in the universe was more energetic (hotter) than any matter we have observed in an experiment. Before this time, we have no experiments that tell us how matter behaves at these temperatures. We must rely on theories, tested at lower temperatures and extrapolated.

There is a very important early time called the *Planck time*, about 10^{-43} seconds after the classical beginning. Before this time, not only are we well beyond any experimental data, but our physical *theories* cease to predict. That is, they can't even be *extrapolated* before this time. This is because we don't have a theory that combines Einstein's theory of gravity with the discoveries of quantum mechanics.

Fortunately, we don't need to probe all the way back to the very beginning to understand the forging of elements in the universe. As mentioned above, the most tightly bound nucleus is that of iron. In practical terms, this means that if we want to reach into the nucleus and pull out a proton, this nucleus charges the highest energy price. If we want to completely unbind a nucleus, not just pull out *one* proton, the most stable nucleus is lead.

As you can probably imagine, the energy needed to completely pulverize a lead nucleus is pretty substantial. But, as we keep stressing, the early universe was very energetic! Before about 10^{-7} seconds after the beginning, *every* particle in the universe had enough energy to break a lead nucleus into its constituent protons and neutrons. The periodic table didn't stand a chance.

This gives the nuclear oven of the big bang unambiguous starting conditions. Even if, by some mysterious process, the universe had some weird mix of elements before 10^{-7} seconds, it doesn't matter. In a moment, all would be protons and neutrons.

In fact, at these very early times, even protons and neutrons are smeared out into a plasma of their constituent quarks. The universe is a sea of elementary particles: electrons, quarks, and their antiparticles fiercely crashing into hot radiation and each other. Every so often, three quarks meet and bind to form a proton or a neutron, lasting infinitesimally before high-energy radiation wrenches them apart again.

As the universe expands and cools, the collisions became less energetic, and protons and neutrons come to live for longer periods of time. (*Longer* is a comparative term, as the universe is only 10^{-6} seconds old!) The processes that make protons and neutrons in the universe have a slight preference for protons because they are lighter. Remember Einstein's famous $E = mc^2$; protons have slightly less mass, and so are slightly cheaper (in terms of energy) to produce. So, as the universe cools, we have a universe that is filled with electrons, protons, (somewhat fewer) neutrons, and photons. Plus some neutrinos, dark matter, and dark energy – more on that in Chapter 10.

An important principle is at work here: if you want to make something in the universe, whether a proton or a lead nucleus, you'll need to wait until things cool off enough for it to survive the ambient temperature.

Our focus here is on the elements, so we ask: when is the universe cool enough for nuclei to survive? For most of the periodic table, this would be between 1 and 10 seconds after the big bang. But we've got a problem: most of the periodic table doesn't exist yet. We have to make small nuclei before we can make big ones. And the smallest stable nucleus is that of our old friend *deuterium*: one proton and one neutron.

Why is this a problem? Because deuterium is more fragile than most other elements: it takes about four times less energy to break a deuterium nucleus than any other. So, we have to wait a whole *minute* before the universe is cool enough that appreciable quantities of deuterium can build up. This delay is known as the *deuterium bottleneck*.

But now the universe can finally make some progress! Protons join with neutrons to form deuterium. Because there are more protons than neutrons, they don't all pair off in this way – there are left-over protons. After this, and very quickly, deuterium and more protons combine to form helium-3 (two protons and a neutron) and then these combine into ordinary helium (known as helium-4, two protons and two neutrons). It seems that the universe is on its way to making a whole host of elements, including the carbon and oxygen essential to you and me.

But the forces of nature are full of surprises. The next reaction has protons and helium-4 to work with. The options are:

- Proton + proton makes deuterium. This won't work, because it requires a very slow weak force interaction.
- Helium-4 + proton makes lithium-5 (3 protons, 2 neutrons). This won't work, because lithium-5 is unstable. It will fall apart before any further reactions can use it.
- Helium + helium makes beryllium-8 (4 protons, 4 neutrons). This won't work either, because beryllium-8 is also unstable.

Inside stars, the next reaction that makes something stable is, in fact, helium + helium + helium makes carbon. But that reaction, because it involves three nuclei coming together, requires high heat and density. Stars can always collapse a bit more and heat up, but the universe, having had to wait a minute for deuterium to bind, is now too cool. There was a time when carbon could be formed, but it has passed, and it's not coming back. The universe just keeps cooling, and carbon has missed its chance.

After only a few minutes of existence, the universe is too cool for nuclear reactions. The repulsive push of nuclei due to their positive electric charge keeps them apart, and *big bang nucleosynthesis* is over!

Computing the Big Bang

It's a nice story, but to make predictions, we need more than words. We need to compute the big bang, to calculate the outcome of the physical processes in these earliest moments of the universe and see what we would expect from a few minutes of cosmic cooking.

The first ingredient is cosmic expansion. As we have seen previously, this depends on the stuff in the universe. During the first few minutes, when the cosmic oven is blazing, the energy of the universe is dominated by *radiation*. This results in a decelerating expansion.

The second ingredient is an understanding of nuclear reactions. As it happens, the physics we need was being investigated around the same time as cosmologists were digging into the details of the big bang.

The discovery of the nucleus of the atom came from one of the scientific greats of twentieth century physics, Ernest Rutherford. At the end of the 1800s, scientists had worked out that atoms are made of smaller pieces. Inside the atom, somewhere, was the tiny electron with its negative charge, and there had to be a corresponding particle in there somewhere, known as the proton, balancing out the electric charge. But just how were these particles distributed within each atom?

Perhaps they were all just mixed together, with the electrons like plums in a positively charged pudding. This idea was first put forward by J. J. Thomson, and Rutherford thought of a clever way to test it. He would bombard atoms with alpha particles, which are chunks of positively charged mass that are spat out of certain radioactive decays. The target was chosen to be a thin sheet of gold atoms, as these are heavy atoms whose pudding must be packed with plums.

The idea was simple: as the alpha particles speed towards the sheet, they see positive and negative charges, and so there will be no overall electric attraction or repulsion. Once inside an atom, an alpha particle will see the pudding of positive charge and plums of negative charge. The pull of the electrons and the push of the pudding will change the path of the alpha particle, but because there are many pushes and pulls, we expect only a small deflection.

Rutherford put two young researchers on the experiment, Geiger (of *Geiger counter* fame) and Marsden. The laborious work of watching for deflected alpha particles was done by eye in a darkened room. A small flash of light indicated that an alpha particle had collided with the fluorescent film. Progress was slow, until a frustrated Rutherford exasperatedly asked Geiger and Marsden to move the detectors 180° around the experiment and see if any of the alpha particles had bounced back.

If the plum pudding model were correct, the number of particles repelled back the way they came would be effectively zero. But Geiger and Marsden did the experiment anyway, and were startled to find alpha particles bouncing back at them! Rutherford famously said that it was "as if you fired a 15-inch shell at a piece of tissue paper and it came back and hit you".[14]

The implications of this experiment are startling. The positive charge in an atom, the charge responsible for repelling the alpha particle, could not be smoothly distributed as in the plum pudding model. It must somehow be concentrated, packed into a central piece: the atomic nucleus. The electrons cannot be inside the positive nucleus – otherwise, it would be neutral and not repel the alpha particles – and so must be orbiting at a relatively large distance from the centre. The modern picture of the atom was starting to emerge.

The discovery of the atomic nucleus triggered a rush of the-oretical and experimental breakthroughs. By 1932, it was clear that the elemental identity of a nucleus depends on the number of protons it contains, but that, in addition, nuclei must contain an electrically neutral particle, dubbed the neutron.

Careful analysis of the elements allowed scientists to weigh the nucleus of each element, revealing something quite striking. A helium nucleus comprises two protons and two neutrons, and we know the masses of these individual components exceedingly well. In atomic mass units (amu),[15] the proton has a mass of 1.007276245 amu, and the neutron is 1.008664956 amu, so the mass of two protons and two neutrons is 4.031882402 amu. However, the mass of the nucleus of a helium atom is 4.001606466, three quarters of a percent lighter!

Through the *fusion* of protons and neutrons into helium, some mass has gone missing! From Einstein's $E = mc^2$, we know mass can be converted into energy. When hydrogen is fused into helium, large amounts of energy are released. It dawned on scientists that this was the Sun's source of power, turning mil-lions of tonnes of hydrogen into helium every second.

To understand nuclear processes, whether in the Sun or in a reactor on Earth, physicists set about measuring and calculating how reaction and decay rates depended on the density and speed of protons, neutrons, and nuclei. Scientists now know these properties extremely well, allowing us to predict how to design a nuclear power plant, or calculate the yield of a nuclear weapon. One of the happy consequences of understanding nuclear reactions in detail is that countries such as the USA and Russia no longer need to build and test new designs of nuclear weapons.[16]

With this new understanding of nuclear physics, in the 1940s cosmologists set about calculating the output of the big bang. In Chapter 5, we met Alpher, Herman, and Gamow, cosmologists who investigated heat and light in the early uni-verse. But in those early days, they faced a lot of unknowns. Which nuclear reaction paths were the most important? Where were the bottlenecks in element production? And what elemental abundances should we expect at the end of this initial burst of nucleosynthesis? While each nuclear species obeys relatively simple equations, the entire network of calculations is quite formidable.[17]

In the 1950s and 1960s, great strides were made by translating the problem into computer code, written in the wonderful Fortran computing language. (We may appear to be biased, having written the core code of our PhDs in Fortran. But actually, it is an objective fact that Fortran is the greatest computer language.) The results were the first robust predictions for the expected abundances of the primordial elements.

Your phone is capable of running the same calculation in a fraction of the time. In fact, you can run your own big bang simulation on your computer. We heartily recommend a code known as AlterBBN.[18] It's nicely documented, including published papers that lay out all the equations and how they are solved. It's open source, so you can pick over all the details, from the expansion of the universe to all the relevant nuclear reaction rates. You can even tinker with the equations, the contents of the universe, or the reaction rates, to consider any universe you want. If you don't wish to compile this free code, you can use the web-based interface at bigbangonline.org, though the equations there are a little more hidden. The results are the same.

The Great Cosmic Bake-Off

To calculate for our universe, one important parameter needs to be input. It's the *baryon-to-photon ratio*, which gives the ratio of the number of ordinary particles of matter in the universe (electrons, protons, neutrons) to the number of photons in the CMB. When these calculations were first undertaken, this was a major unknown in the equations. Remember, this was the 1960s, about the same time as the CMB was being discovered. It would be decades before other observational data – including surveys of galaxies, more detailed observations of the CMB and the Lyman alpha forest (Chapter 6) – would measure the cosmic density of ordinary matter with the required accuracy. But now we have measured this number, independently of big bang nucleosynthesis data: $(6.10 \pm 0.04) \times 10^{-10}$. In other words, there are roughly 1.6 billion CMB photons for every proton, neutron, and electron.

Finally, the answer. The big bang's observed and predicted elemental output is summarized in the table below for deuterium, helium-3, helium-4, and lithium-7 (3 protons, 4 neutrons). The uncertainty in the baryon-to-photon ratio leads to a small uncertainty in the predictions. Taking this into account, we can state the big bang's predictions for the primordial element abundance of the universe:[19]

	Observed	Predicted
Helium mass fraction	0.2534 ± 0.0083	0.2463 ± 0.0003
Deuterium fraction (per 10^5 protons)	2.569 ± 0.027 (Fields et al.[10]) 2.527 ± 0.030 (Cooke et al.[11])	2.67 ± 0.09
Helium-3 fraction (per 10^5 protons)	1.1 ± 0.2	1.05 ± 0.03
Lithium-7 fraction (per 10^{10} protons)	1.58 ± 0.31	4.89 ± 0.4

Remember: the crucial question is whether the uncertainty regions overlap, which tells us that observation and prediction are consistent. To help visualize the numbers, we have plotted the ranges relative to the observed value for each nuclear species in Figure 7.1.

Let's run through them.

- Helium is successful. The very thin grey prediction is inside the observed black box. This is a remarkable success for the big bang theory.
- Deuterium is also consistent. One measurement (Fields et al.) agrees with the prediction, and the other (Cooke et al.) is only slightly outside. This is another success for the big bang theory.
- Helium-3 is successful, but don't get too excited. Its measurement is fraught with uncertainty – we're not completely sure about all the production and destruction

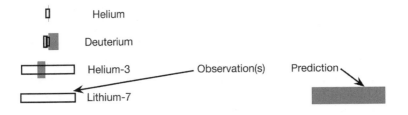

Figure 7.1. Comparing prediction and observation for big bang nucleosynthesis. The observed values are shown as black outlines, and the predicted values as grey boxes. Each range of values is shown relative to the observed measurement (from Fields et al.,[10] in the case of the two deuterium measurements).

processes of helium-3 in stars. It's a tentative pass-mark for the big bang.

- Lithium-7 ... um. Oh dear. The prediction is three times higher than the observed value. Given that the observed fraction is about one part in 10 billion, we're in the right ball park. But it's still not right. We'll discuss this case further when we collect this and other problems for the big bang in Chapter 10.

All in all, the scorecard for the big bang is impressive. The cosmologist Paul Davies, in his book *The Last Three Minutes*, recalls attending a cosmology lecture in his student days, during which the lecturer claimed that "some theoreticians have given an account of the chemical composition of the universe based on the nuclear processes that occurred during the first three minutes after the big bang." The entire audience, he tells us, "laughed uproariously" at this preposterous idea.

And yet, it is remarkably successful, lithium notwithstanding. In the time it takes to cook an egg, the universe cooked a quarter of its mass into the light elements.

A Contender Falls: The Electric Universe

Fred Hoyle's name is wrapped up in many aspects of this chapter, being one of the key astronomers who linked the physics of the nucleus to the physics of stars and the cosmos. But in Chapter 5, we saw that he was not a fan of the big bang cosmological model, pushing hard for his preferred steady-state universe in which matter was continually created over time, resulting in an expanding but on-average unchanging universe.

In spite of this, he was one of the physicists who first calculated the predictions of the big bang model. This illustrates a vital point: your idea about the universe should be so unambiguous and mathematically precise that any physicist – whatever they think of the theory – can calculate its predictions. That's how you know that your idea isn't just a vague dream that only you understand.

The observed abundance of elements in the cosmos is yet another nail in the coffin for the steady-state model. With only stars to make heavy elements, it cannot explain a 25% helium universe, nor can it explain the presence of *any* deuterium in the wider universe.

The cosmos's elemental abundances are one of the central pillars of the big bang model, providing key evidence that the universe had a hot, dense beginning. But what about other contenders?

Some physicists have proposed cosmological models in which gravity is not the dominant force in the universe. Rather, given that much of the universe is an ionized plasma, electricity and magnetism rule cosmic motion and evolution. This idea has a good pedigree, coming from Hannes Alfven, a Nobel Prize winner for his work on plasmas. His "plasma cosmology" or "electric universe" has a significant following among a small community of physicists, who are generally not cosmologists.

Plasma cosmology, however, is not a single, mathematically formulated, physical theory, but more of a catch-all for a collection of (sometimes wild) ideas. There is no fiery beginning in these models, so the story of the elements in plasma cosmology is going to be very different from the big bang. What do they say about the cosmic abundance?

To add a twist to the tail, plasma cosmologies deny that stars are powered by nuclear reactions at their core. They claim that the surface of the star interacts with large-scale electric and magnetic fields in the wider universe. This interaction heats the surface, causing nuclear reactions to fire up in the *outer* layers of the star, not the core. Furthermore, stars are not considered to be cosmological slow cookers, whose properties are dictated by their mass. They are controlled by their local electromagnetic environment, making their evolution terribly messy.

The problem with these pronouncements is that they are just words. They lack mathematical substance. The proponents of plasma cosmologies have not made predictions that can be held up to observations. For plasma cosmologies to be taken seriously as an alternative to the big bang, they need to put up or shut up. Go calculate the synthesis of elements, following the evolution of electric stars through their complex history. Explain your equations and methodology well enough that any physicist could do the calculation themselves. Tell us the elemental abundance of the plasma cosmos. And then see if you got it right.

Until plasma cosmology can do that, and provide new observational tests to distinguish it from the big bang model, it will remain in the shadows.

8 RIPPLES IN THE NIGHT SKY

In this chapter, we are going to take a slight diversion. So far, we have taken a peek at our best observations of the sky, how they are interpreted by the big bang model, and how other models have failed to account for our best evidence. In the last 100 years, the big bang theory has faced a number of problems. To understand this, we need to travel back to the early universe and think about perfect physics.

The Missing Magnetic Monopole Mystery

When physicists talk about the deepest laws of the universe, they place a lot of emphasis on the word *symmetry*. Since the middle of the last century, writes Nobel Laureate David Gross, "symmetry has been the most dominant concept in the exploration and formulation of the fundamental laws of physics."[1] But just what do scientists mean by symmetry?

You probably have a geometric idea of symmetry: if someone says, "this vase is symmetrical", it means (something like) the right side of the vase looks like a reflection of the left side. There are many kinds of geometric symmetry, depending on which part looks like which other part. A sphere is the perfect example: every part looks like every other part. The acerbic astronomer Fritz Zwicky would refer to certain colleagues as "spherical bastards", meaning that they were bastards no matter how you looked at them.

Symmetry in fundamental physics has a similar meaning, namely that something looks the same when you look at it differently. But these symmetries more often relate to equations than to geometric shapes. We transform quantities rather than manipulating shapes, and find that the equation that relates them looks the same. This is the kind of symmetry that we need to consider.

The physics of the everyday world often doesn't appear very symmetrical. Consider electricity and magnetism, which have been known for millennia, and treated as quite separate physical phenomena. Beginning in the early 1800s, it was realized that they are somehow connected: electricity can produce magnetism and magnetism can produce electricity. This interconnection was given its deepest expression in 1861 by the great physicist James Clerk Maxwell, who showed that one set of equations governed all electric and magnetic phenomena. History doesn't record who first printed these equations on a t-shirt.

The entirety of our understanding of classical electric and magnetic fields (that is, where we don't have to worry about the effects of quantum mechanics) is wrapped up in Maxwell's equations, but there is something uncomfortable about them. There seems to be a nice symmetry between the electric and magnetic fields: changing an electric field produces a magnetic field, and changing a magnetic field produces an electric field. But there is one noticeable difference. Static electric fields are created by electric charges, but there are no magnetic charges that create static magnetic fields. This leaves Maxwell's equations looking less than perfectly symmetric. If the universe possessed a particle with magnetic charge, or a *magnetic monopole* as it is known, then the equations would be beautifully symmetric: magnetic fields would emanate from these charges, and electric fields would be created by currents of magnetic charge. But, alas, all searches for a magnetic cousin have drawn a blank.

Now, maybe magnetic monopoles just don't exist. But the symmetry calling out from Maxwell's equations is so alluring, almost too perfect to ignore. Many theoretical physicists have looked for a different kind of answer: magnetic monopoles exist, but their rarity is something peculiar to our universe, rather than something deep in the laws of nature. Something happened in the history of our universe to create the observed asymmetry between magnetic and electric charges. And the prime suspect is the early universe.

As we have seen, in the initial stages, our universe was extraordinarily hot, with particles and radiation bouncing around with huge amounts of energy. In many ways, this makes the universe a much simpler place than today; for one thing, there are no nuclei or atoms, so no nuclear or chemical reactions to worry about. But there is another, deeper reason why the early universe is quite simple.

This realization comes from an examination of three of the fundamental forces of nature:

- the strong nuclear force, which holds together the nuclei of atoms;
- the weak nuclear force, which is responsible for certain types of radioactivity;
- electromagnetism, which is described by Maxwell's equations.

As we consider environments where particles have more and more energy, such as in the early universe, these forces begin to show the same kind of interconnection we saw for electricity and magnetism. The distinction between these forces starts to blur, and there is effectively only one force acting: a superforce! Only after the universe cooled did the superforce fracture into the forces that we see today.

Gravity is thought to have been united with the other forces into the superforce just after the big bang. However, because we don't know yet how to fit gravity together with the other forces in a unified way, gravity stands apart, thumbing its nose. This ultimate unification of forces remains a goal/dream of modern physics.

The beautiful symmetry of the early universe suggests that we should have such a symmetry in Maxwell's equations. And if magnetic monopoles can exist at all, then we would expect them to have been created in the intense energy of the early universe, in the same way that electrons and quarks were forged. But they aren't to be found.

In light of this incorrect conclusion, we need to identify all the assumptions in the model, because at least one of them needs to go in the garbage. We are reluctant to throw away the electromagnetic symmetry, or to jerry-rig the superforce so that it doesn't make monopoles. Is there another option?

The Inflationary Universe

In the early 1980s, Alan Guth, now at the Massachusetts Institute of Technology, was a young theoretical physicist who was worried by those missing magnetic monopoles. In particular, he was looking closely at the cooking of magnetic monopoles in the early universe.

To create any particle, there is a cost in energy. Einstein's $E = mc^2$ tells us, for example, that making a proton will cost you 1836 times more than making an electron, because a proton is 1836 times heavier. So, as the early universe cools, there comes a time when collisions between its particles are no longer able to make a given particle. The heavier the particle, the earlier the universe ceases producing it. For a very heavy magnetic monopole, the production line stops very early on.

So, there is a *window* in the early universe, between the time when magnetic monopoles are made and the time when lower mass particles such as quarks and electrons are made. It's a fraction of a second, but long enough to rethink our assumptions about what would happen.

Guth wondered if something unexpected could have happened in this window. Specifically, what if the universe expanded much faster than expected, accelerating rather than decelerating? This kind of expansion is called *inflation*. It needs a cause, in the form of an energy field with some particular properties. We call this field the *inflaton*, so that when physicists write songs about particles like protons, electrons, photons, and neutrons, it's easy to add this new field into the rhyming scheme; for example, a particle with mass 925 is called a dolly-parton.

The inflaton field is a source of immense anti-gravity, causing a very short but very intense burst of accelerating expansion.

Between 10^{-34} and 10^{-32} seconds after the big bang, inflation rapidly dilutes the contents of the universe, leaving the universe full of inflaton energy but almost completely devoid of particles. As inflation ends, it dumps its energy back into ordinary radiation and particles, and thereafter the history of our universe plays out as usual.

Here's the trick. Firstly, make sure that inflation happens after the monopoles are created. Then, they are rapidly diluted by the expansion, leaving approximately one monopole in the observable universe; no wonder scientists are having a difficult time finding one. Secondly, when the inflaton dumps its energy back into ordinary matter, make sure it's not hot enough to re-make the magnetic monopoles.

After inflation, the universe continues cooling as if nothing had happened. The universe would be too cool to create more magnetic monopoles, but still hot enough to fill the universe with the quarks and electrons that eventually formed atoms, then stars, planets, and people.

Obviously, there are a few pieces of this story that are missing. What is the inflaton field? What caused inflation to start, that is, why did the inflaton field take control of the expansion of the universe? And why did it later relinquish control, causing inflation to end? And by what process did the energy return to ordinary matter and radiation?

Many physicists have pondered these questions since the early 1980s: Alan Guth, Alexei Starobinsky, Demosthenes Kazanas, Andrei Linde, Paul Steinhardt, Andreas Albrecht, Stephen Hawking, Viatcheslav Mukhanov, So-Young Pi, and more. Many, many trees have been sacrificed as ideas have been transformed into manuscripts; thankfully, many scientific journals have morphed into electronic-only journals. Four decades later, we have a number of candidate models for inflation that do all the work required to solve the magnetic monopole problem, but we still don't truly grasp the physical processes involved. What we need are more predictions from inflationary theories, so we can sort the winners from the losers. Like all theories that capture the attention of physicists, inflation has a few more tricks up its sleeve.

Seeing to the Horizon

A problem lurks within the CMB. As we saw in Chapter 5, when the CMB was discovered in the 1960s, it was apparent that this unexpected light was coming uniformly from all parts of the sky. This uniformity became increasingly impressive as observations improved. We now know that wherever we look in the sky, we see the same temperature of 2.725 degrees above absolute zero, to within 0.001%. As cosmologists considered how the universe managed to be so uniform, they found a puzzle.

To illustrate the problem, let's take a look at human history. Tens of thousands of years ago, humans were spread across the globe, from Australia to the tips of the Americas. As people thought new thoughts and invented new inventions, these ideas took a long time to spread from region to region. When the usual communication medium is word of mouth, ideas spread as fast as walking. An innovation discovered at the tip of South America, such as a new way to light a fire or a wondrous new metal, would slowly permeate from the point of discovery. Sometimes ideas would peter out after travelling a short distance or be obstructed by mountains and oceans. In this way, culture, language, art, and technology vary in different parts of the world.

But over the last few thousand years, and especially over the last few hundred years, the speed of communication has rapidly increased, firstly through global travel and more recently through electronic communication. The human race is sharing ideas faster than ever, with innovations spreading effortlessly all over the globe.

While great for the world economy, with businesses able to sell their products to markets numbered in the billions, this globalization of innovation isn't all good news. As you travel the planet, you start to get déjà vu: in many large cities, you'll find McDonalds and Pizza Hut and Heineken. On a walk down the high street, you will find the same shops, selling international fashion or media. Most cinemas are "Now Showing" the same movies. And is there a place on the planet where you cannot find famous fizzy brown American lolly water? If this book makes it to those parts, please let us know.

Everywhere starts to look the same.

But suppose that the ancient geographer Pomponius Mela, writing in about AD 43, had been correct about Earth. He believed that the equator is too hot for humans to survive. No one, and no message, can cross between the hemispheres. This led to centuries of speculation about whether the Southern Hemisphere had any occupants at all, "men who walk with their feet opposite ours".[2] In this case, there would be no way for the ideas of the north to permeate to the south, and vice versa. The streets, shops, and culture of one hemisphere would be quite distinct from the other.

This idea is explored by science-fiction writer Harry Harrison in his book *West of Eden*. In this parallel universe, the dinosaurs of Earth were not wiped out 65 million years ago, but instead have evolved into cultured and humanoid reptiles. However, there is one part of the planet where these reptiles never lived, allowing human-like creatures to evolve in isolation. When the lizards and humans finally meet, the results are catastrophic.

What does this have to do with cosmology? It illustrates an important principle about the flow and mixing of energy. Imagine we have two sealed rooms, one containing a cold gas, where molecules move sluggishly, and one containing a warm gas, with molecules zipping about. Now we open a door connecting the two rooms, and the gases start to mix. When a slow-moving particle collides with a fast one, they tend to share their energy, resulting in two particles moving at intermediate speeds. As the gases from the rooms mix, the zippy particles from the hot room will tend to slow down, and sluggish ones speed up. If we wait long enough, the gas in the two rooms will reach identical temperature. The key point is that, in this instance, identical temperature was the result of *thermal contact*.

Now let's think about the birth of the universe. We don't know the process that brought the universe into being, but our current physical theories tell us that the epoch just after creation would be extremely hot and extremely dense, conditions that demand that we cannot neglect the impact of gravity and quantum mechanics. The theory of quantum mechanics is interpreted by many (though not all) physicists to imply that the universe has a fundamental randomness to it. The universe today doesn't completely dictate what the universe must be like tomorrow; there is an element of chance, too. So perhaps, in the very early universe, fluctuating fields of energy produced varying temperatures in separate patches of the universe, a bit hotter over here and a bit cooler over there.

In that case, the very early universe was like the rooms in the example above, with patches of hot gas alongside patches of cool gas. Because there is no barrier between the patches, we would expect energy to flow, from hot to cool, levelling out the initial temperature fluctuations. The universe would soon come to the same temperature, just as we see in the CMB. All is good, right?

Alas, no. We forgot that the universe is expanding. We assumed that energy would flow from hot patches of gas to cool patches of gas, but the fastest this heat could be transmitted is at the speed of light; nothing can move faster. Is it fast enough?

Imagine we are looking at two patches of the CMB on opposite sides of the sky. This microwave emission comes from a time when the universe was about 380,000 years old; the patches of gas that we are seeing are now on opposite sides of the universe, separated by 90 billion light years. Today, even if the universe's expansion stood still, it would take 90 billion years for energy to flow from one patch to the other. But there has only been 13.8 billion years since the big bang. The expansion of space has separated these two patches by such a great distance that equalization of temperature today is seemingly impossible.

Of course, it is not the separation *today* that matters. The universe was smaller in the past – that helps. On the other hand, less time has passed since the big bang – that makes things worse. As we wind the clock back, is it easier or more difficult for these patches to be in thermal contact? Does the expansion of the universe make its matter more or less connected?

Using the equations from Einstein's relativity, this is straightforward to calculate. A million years after the big bang, the patches were separated by 150 million light years. A thousand years after the big bang, they were separated by 3 million light years. A year after the big bang, they were separated by a hundred thousand light years. One second after the big bang, they were separated by 18 light years (over 500 million light seconds).

The problem only gets worse. No matter how far back we push, there *never* was a time when the two patches, which now appear on opposite sides of the sky, were close enough to achieve a uniform temperature. The two patches are, and always have been, beyond each other's *horizon*.

In fact, the patches don't have to be on opposite sides of the sky. Put on your favourite pair of microwave-detecting goggles and look up at the CMB. Point at one part of the night sky, and then another part just 1 degree away, which is about twice the size of the full Moon. Follow your pointing fingers out into space and back in time to two patches of hot gas, 380,000 years after the big bang. Those two patches of gas have never been in contact with each other. That is, light from one region cannot have travelled to the other. They are beyond each other's horizon. So, whatever set the temperature of the first patch cannot have influenced the temperature of the second patch. We might expect, then, that the temperature of the CMB radiation would be blotchy across the sky, some patches hotter, some patches cooler. But this is not what we see. We see the same temperature in all directions. How? This is known as the *horizon problem*.

Before we look at how inflation addresses this issue, we need to hit the brakes. We should take a close look at an assumption underneath this argument:

1. The patches of gas that we see in the CMB have the same temperature.
2. If the patches of gas that we see in the CMB had been in thermal contact, we would expect them to have the same temperature.
3. If the patches of gas that we see in the CMB had not been in thermal contact, we would not expect them to have the same temperature.

The first statement is true – we observe this directly in the night sky. The second statement is true, so long as they are in contact for a sufficient amount of time. This is basic physics – ice melts and coffee cools and all that. However, the third statement, while being crucial to the horizon problem, is a bit of a guess. (It doesn't follow from the second statement – that's a logical fallacy.) If the explanation of the uniform temperature is not ordinary thermal contact, then we're left asking: what would we expect the universe to be like at its very beginning? And that's a very hard question, because we don't have a theory that tells us what the very beginning of the universe was like. Maybe, when we get a Theory of Everything, we'll realize that uniform universes are the norm. We just don't know.

Nevertheless, we can ask the question: is there a change we can make to the story of our universe – from within the physics that we actually know – that would explain the uniform temperature of the CMB? Cosmic inflation is just such an idea.

Enter Inflation

So, how does inflation rescue the situation? The key here is that the calculation about what regions of the universe are and aren't connected depends on how the universe expands. A crucial step in creating the horizon problem is the assumption that, in the earliest stages of our universe's history, the expansion was decelerating. What happens if we put an inflationary era, with its rapidly accelerating expansion, into the early universe?

Before inflation, let's assume that the universe was a sea of matter at different temperatures, some cooler, some hotter. (It wouldn't be much of an achievement if we started by assuming uniform temperature.) Due to the usual decelerating expansion, these regions are disconnected from each other, and energy cannot flow across the universe and equalize the temperature everywhere.

But suddenly there is the burst of inflation, lasting a tiny fraction of time, but swelling the universe by an enormous factor. Each microscopic patch of hotter or colder matter gets blown up. Our observable universe is fully contained in one patch. In other words, a patch of the early universe small enough to reach the same temperature is now bigger than our observable universe. So, our microwave telescopes see an equal temperature in all directions.

This implies that out there – deep, deep in the universe, in regions far beyond our cosmic horizon – distant alien worlds will see a hotter or cooler CMB than we do, sitting within a different patch of matter that was disconnected from us during inflation. They too will see a uniform temperature across their microwave sky, and it is interesting to ponder whether their alien cosmologists are also wondering why this is the case. Perhaps they will conclude that their universal expansion needs a period of inflation to explain their sky.

Molehills out of Mountains

Like any good second-hand car dealer, inflation cries "but there's more!" Inflation has other consequences, explaining another bonus feature of the universe. Inflation explains why the universe appears to be so *flat*!

To understand what a cosmologist means by *flat* in this context, we have to dig a little into the mathematics underlying cosmology. In Einstein's description of gravity, the energy in the universe affects its spacetime geometry. You hopefully remember some geometry from school: distances and angles, triangles and pyramids. This geometry has been passed down to you through over two thousand years of mathematicians, and especially from the famous ancient Greek geometer Euclid, and his classic textbook *The Elements*.

The geometry of Euclid is the geometry of flat surfaces. When you draw a triangle on a flat piece of paper, the internal angles add up to 180°. If it's a right-angled triangle, then Pythagoras's theorem is true. When you draw a circle on a flat piece of paper, the circumference of a circle is equal to the radius multiplied by 2π. We can extend this geometry into three dimensions; for example, the surface area of a sphere is its radius squared multiplied by 4π. As the laws of Euclid hold, we call these spaces *Euclidean*.

But not all surfaces are flat like a piece of paper. If we started drawing lines and triangles and circles on a curved piece of paper, Euclid's laws would not hold. A familiar example of this is the surface of the Earth, which we can think of as the surface of a sphere. When airline pilots are taught how to navigate a plane, they learn a different kind of geometry. For example, suppose Lois and Clark, our explorers from Chapter 3, want to fly some enterprising tourists from Quito, Ecuador (which is approximately on Earth's equator), on a day trip to the South Pole. The flight plan is as follows.

- Set out from the equator and fly directly south to the South Pole, enjoying the sights.
- At the South Pole, take a sharp right-angled turn to the left and fly towards the equator.
- Arrive at the equator a quarter of the way around the Earth from Ecuador, specifically somewhere over Equatorial Guinea. Take another sharp right-angled turn to the left and fly along the equator back to Ecuador.
- On arriving back at Ecuador, make a sharp right-angled turn to the left and land, so that the plane is pointing to the south, ready for the next trip.

Think carefully about what just happened. The plane flew in straight lines relative to the surface of Earth, not steering to the left or the right. It arrived back where it began, so it traced out three sides of a triangle. And yet, the sum of the internal angles is $90° + 90° + 90° = 270°$. Try this on a flat piece of paper: it can't happen.

Similarly, Pythagoras theorem doesn't work on the surface of Earth. The actual distance between two points will differ from the usual formula, and how wrong you are depends on where you are on the Earth's surface.[3] If airline pilots, and those who design their navigation systems, didn't take this different geometry into account, every plane would miss its destination.

To accurately navigate a curved surface, a new geometry was needed, a *non-Euclidean* geometry. This was developed in the

1800s, through the works of mathematical greats such as Carl Friedrich Gauss, Nikolai Lobachevsky, Bernhard Riemann, and Elwin Christoffel.

Let's think about the surface of the sphere a bit more. Nikolai, an intrepid explorer (is there any other kind?), has reached the North Pole, and wants to claim a circular area around the Pole for his homeland. In particular, he wants to build a fence to keep everyone else out. How long will the fence be? He places a stake into the snow at the North Pole and ties a rope of known length to it. He walks directly away from the stake until the rope is taut, turns 90 degrees and starts walking, always keeping the rope taut, and counting every step. Eventually Nikolai will return to where he started, having traced out a circle. He knows the radius of the circle (the length of the rope), and how far he walked around its circumference.

If the Earth were flat like a piece of paper, the circumference would be 2π times the radius. But on the curved surface of the Earth, we get a different answer. The circumference is smaller than we expect on a flat surface. If the rope is 10 km long, the difference is only about 3 cm. If the rope is 100 km long, Nikolai's fence will be 26 m shorter than expected. And if Nikolai tries to claim most of the North Pole ice cap with a 1,000-km rope, his fence will be about 26 km shorter.

If you have a globe to hand, you can try this experiment yourself. Lay a piece of string along a line of latitude, following a circle right around the model Earth. Then take another piece of string and place it along a line of longitude, from the point you started to the nearest pole. Measure the length of each piece of string to see if the second string is 2π times the first one.

We must also change our assumptions about parallel lines. On a flat sheet of paper, if two separate lines point in the same direction, they will never meet. But recall from Chapter 3 that Lois and Clark, starting a certain distance apart on the equator and travelling south, meet at the South Pole, even though they travelled in locally straight lines, not steering left or right. In fact, any direction you choose will result in your paths crossing eventually.

Mathematically, we say that the Earth's surface, on which the internal angles of triangles add up to *more than* 180°, is *positively* curved. Surfaces can also be *negatively* curved: if an ant walked around a saddle, drawing a triangle as it went, the internal angles would add up to *less than* 180°.

These are two-dimensional examples, but we can mathematically move them into three dimensions: circles become spheres, areas become volumes. The downside is that it's impossible for us to imagine. We thought about a curved two-dimensional surface by embedding it in three dimensions. But unless you can imagine the "surface" of a sphere embedded in *four* dimensions, this strategy isn't going to work.

Nevertheless, the implications are straightforward. In a curved three-dimensional space, parallel paths may be multiple (negatively curved) or impossible (positively curved). The surface area of a sphere of radius r is not the Euclidean answer of $4\pi r^2$. The internal angles of triangles don't add up to exactly 180°. The mismatch depends on the size of the sphere or triangle, relative to the curvature of the space.

How Curved Is My Universe?

In the decades after mathematicians worked out these properties of curved spaces, the implications dawned on physicists. Your backyard appears to be flat, but in fact is just a small part of our spherical Earth. What if three-dimensional space is like your backyard – locally nearly flat, but curved on much larger scales?

The idea of curved spaces is most often associated with Einstein in the early twentieth century, but in fact the idea had entered public consciousness in the nineteenth century. Thirty-five years before Einstein, in Fyodor Dostoyevsky's 1880 novel *The Brothers Karamazov*, Ivan comments to Alyosha:

> ... *there have been and there still are mathematicians and philosophers, some of them indeed men of extraordinary genius, who doubt whether the whole universe, or, to put it more wildly, all existence was*

created only according to Euclidean geometry and they even dare to dream that two parallel lines which, according to Euclid can never meet on earth, may meet somewhere in infinity. I, my dear chap, have come to the conclusion that if I can't understand even that, then how can I be expected to understand about God?

It was Einstein who connected the geometry of space and time to the phenomenon of gravity. In his theory, as applied to a homogeneous and isotropic universe, space could be flat and Euclidean, but it does not have to be. It could be positively or negatively curved. This sounds esoteric, but it has very observable consequences – you just need a big enough triangle. Cosmologists have devised a battery of tests to measure the geometry of the universe. It's not easy, because as well as being possibly curved, the universe is expanding.

The most accurate tests come, as usual, from the CMB. This light has travelled through over 13 billion years of the universe's history, its path determined (in part) by the geometry of the universe. The CMB is almost perfectly uniform, but on closer examination is awash with tiny ripples, minute deviations in temperature away from the average.

With a lot of hard work, astronomers have been able to calculate the *physical* size of these ripples, by modelling the flow of matter and radiation in the early universe, which can then be compared with the *angular* size of the ripples on the sky. The most detailed observations on the CMB come from the Planck satellite, which can pick out the bumps and wiggles on the microwave sky as small as a sixth of a degree (see Figure 8.1).

The sizes of the ripples are not like the size of a big ruler, with definite ends. Rather, the sizes manifest in the statistical pattern of the bumps and wiggles of the CMB.

To understand the mathematics of this measurement, consider an analogy. You have some coloured bits of paper:

- 10,000 small red pieces, each the size of your thumbnail;
- 5,000 medium yellow pieces, the size of a watch face;
- 1,000 large blue pieces, as big as the palm of your hand.

Figure 8.1. Fluctuations in the temperature of the cosmic microwave background. The darker regions are slightly hotter, and the lighter regions slightly cooler than average, by one part in 100,000. (Based on observations obtained with Planck (www.esa.int/Planck), an ESA science mission with instruments and contributions directly funded by ESA Member States, NASA, and Canada.)

You give the paper to some small children, who are told to spread them around on the floor, leaving no overlaps and no gaps. After letting them loose for 10 minutes, they will have created a modern artwork titled *Why Parents Are Tired*. You take a photograph of the colourful floor and then painstakingly gather up the paper pieces, once again forming piles of red, yellow, and blue pieces. And then, being a good scientist, you bring in the kids and do it all again, multiple times.

When you have collected a large number of photographs, you will see that they are similar, but different. You know that there are the same numbers of blue, yellow, and red pieces of paper in each of the images, so there are average properties that remain the same in each picture. But they are scattered differently, so whether a certain section of floor is red, yellow, or blue changes from image to image.

Suppose we took blurry photographs, so that the edges of the pieces of paper couldn't be seen. There would just be a continuous pattern of colours. We can use statistical tools on each photograph to calculate the typical size of red, yellow, and blue pieces of the pattern, and to calculate the relative abundance of each colour.

The cosmic microwave sky is just like this. We see patterns of hot and cold, bumps and wiggles. Some of the details – such as whether some direction in the sky is hot or cold – are random and would be different if we were looking at the CMB from somewhere else in the universe. But the pattern also has overall statistical properties, similar to the relative numbers of small and large pieces of paper. These are what cosmologists measure and compare to the predictions of our model of the early universe. In particular, from this data, we can measure the curvature of space itself.

And the answer is . . . our universe is so close to flat that we can't tell whether it's actually flat, or very, very, very slightly curved.

What an anticlimax! But what does it mean?

Suppose you tried to measure the curvature of the Earth in your backyard. It's technically possible, but it would require extraordinarily precise instruments. If your instruments aren't accurate enough, your conclusion should *not* be that the Earth is flat, but rather that the Earth is so close to flat (on the scale of your backyard) that we can't tell whether it's actually flat, or very slightly curved.

Science, in comparing nature to theory, must always deal with the uncertainties and errors of all our observations and experiments. No measurement is perfectly precise. Future experiments can only reduce these uncertainties, but they cannot eliminate them.

Sometimes, the conclusion that we want to reach won't depend on the uncertainty, so we can make a definite statement. We don't know the mass of the proton and the mass of the electron to perfect precision, but we know that the proton is heavier. The ratio of their masses is $1836.15267389 \pm 0.00000017$. So, while we are not certain whether the ratio is 1836.15267372 or 1836.15267406, we're quite certain that the ratio is greater than one.

But sometimes the uncertainty matters. There is a quantity in cosmology that measures the curvature of space, known as Ω_k ("Omega k"). If it's equal to zero, then the universe is flat. The best measurement we have tells us that $\Omega_k = 0.001 \pm 0.002$. As you can see, the region of uncertainty lies on both sides of $\Omega_k = 0$. The only conclusion we can draw is that the universe is so close to flat that we can't tell whether it's actually flat, or very slightly curved. The data just won't let us say anything more.

But why is it nearly flat? This question has some extra urgency because of the following fact about how space expands. If the universe is born perfectly flat, it will stay perfectly flat forever. That is, if $\Omega_k = 0$ at the beginning, then it is always zero. But if the universe is born with some curvature, then, on the standard big bang model, that curvature will evolve away from flatness. The number Ω_k will get further and further from $\Omega_k = 0$. So, for our universe to be (so-close-we-can't-tell) flat today, it must have been extremely close to flat in the past. When the universe was forging its elements in the first few minutes, the universe must have been within 1 part in 10^{16} of perfect flatness.

Is that what we'd expect?

Now, we need to hit the brakes again. As above, we've asked a question about what we expect of a newborn universe. And that's a question with some guesswork involved, because we don't have a theory that can predict what the beginning of the universe was like. In fact, some have argued – on the basis of Einstein's theory of gravity and also on purely probabilistic grounds[4] – that flatness is exactly what we'd expect, with complete certainty. So, there's no problem here, no mystery. Cosmologists can just carry on.

Nevertheless, these ideas are controversial. In particular, the conclusion – certainty about flatness – seems awfully strong. As with the horizon problem, we can ask a smaller question: without assuming perfect flatness for our newborn universe, is there a change we can make to the story of our universe – from within the physics that we actually know – that would explain its flatness? Again, cosmic inflation fits the bill.

Stretching Your Space

The key is inflation's very rapid expansion of space. Think about your backyard again: the reason it's hard to measure the Earth's curvature in your backyard is because it's so much smaller than the Earth. The bigger the planet you live on, the harder it would be to measure its curvature in any fixed region.

So, even if our part of the universe were strongly curved at early times, the very rapid expansion due to inflation would almost completely flatten it. The universe would become so big that it would be practically impossible to measure its curvature. Anyone in such a universe, when they tried to make such a measurement, would get our answer: so close to flat that we can't tell whether it's actually flat, or very slightly curved.

In other words, during a period of accelerating expansion, the curvature parameter Ω_k is driven *towards* zero, not away from it. So long as inflation lasts long enough, and stretches the universe to be big enough, the curvature parameter Ω_k will be so close to zero that we can't tell.

This is quite a neat result, and one of the main reasons why inflation has been studied by so many cosmologists. After all that effort, where do we stand? How convincing is the case for cosmic inflation?

Is Inflation a Contender?

So, this is where things get weird. Science at the limits of our knowledge doesn't look like the usual image of science – careful, factual, methodical, objective, widely accepted. Instead, scientists are trying out ideas, making guesses, taking creative leaps, speculating beyond current experiments, and disagreeing at every turn, sometimes with great passion.

In the last 40 years, cosmologists have pointed to four more features of inflationary models that lend them support.

- Inflation provides a mechanism that generates the initial lumps and bumps in the universe (the 1 part in 100,000 we see in the CMB). However, different models predict different amounts of lumpiness.

- An inflationary model will predict the distribution of the sizes of the lumps and bumps in the universe (think of the distribution of red, yellow, and blue paper in the example). This has been confirmed. However, inflation is not the only model that predicts this distribution, which was written down and studied by cosmologists in the 1970s, before inflationary models were invented.
- Inflation predicts that, other than the distribution of sizes, the pattern is as random as possible, technically known as a *Gaussian random field*. But again, randomness is not hard to generate.
- Inflation predicts that, at some level, there will be an extra layer of lumps and bumps due to gravitational waves. To date, this has not been observed. We will discuss this further below.

In light of these points, the cosmology community is split. Inflation is not universally accepted. Here are the cases that each side puts forward.

The "Exploding Universe" Party[5]

The case for inflation goes like this.

- Many grand unified theories predict magnetic monopoles, so we can award a few points to any theory that explains why they aren't commonly observed.
- Without inflation, only one very specific distribution of temperatures when the universe was born – almost uniform – will explain the CMB. Again, solving a plausible problem is worth a few points.
- Unless we are confident that almost all universes are born without curvature, the flatness problem is alive and well. Score a few points for any theory that solves it.
- Inflationary theory explains how the properties of the lumpiness of our universe are generated, and, in particular,

are coherent on scales larger than the distance that light could have travelled on the standard big bang model. More points.

- Let the alternative models of cosmic lumpiness come forth! We will examine them, to see whether they are simpler and do a better explanatory job than inflation. The fact that none have commanded comparable attention from cosmologists speaks to the appeal of inflation. Chalk it up.
- We just need to look harder to see the subtler effects of inflation, such as the imprint of gravitational waves. This is no failure of inflation.

Inflation is the best theory we have of cosmic initial conditions. The choice is between a known solution to plausible problems, and an unknown hope that the problems might go away.

The "We Don't Need No Inflation" Party

Here is the case against inflation.

- We don't know whether monopoles are physically possible because we don't have an experimentally tested grand unified theory. It's easier to say that we don't see monopoles because there are none. So, the monopole problem is a non-problem. No points.
- We don't know what distribution of lumps and bumps we would expect from the birth of the universe. So, the horizon problem is a non-problem. No points.
- We don't know whether to expect a flat universe (or, more strongly, we should expect a flat universe). So, the flatness problem is a non-problem. No points.
- Inflationary models don't uniquely predict the amount of lumpiness, so that's not a win. Maybe half a point for having a mechanism for generating lumpiness.
- The imprint of gravitational waves has not been observed. No points.

In fact, this last bullet point earns inflation some negative points, in the opinion of some, in light of a peculiar non-event a few years ago.

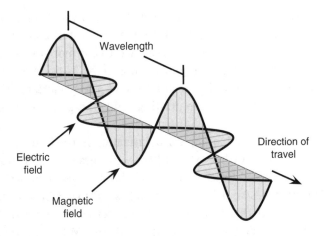

Figure 8.2. An electromagnetic wave, with the electric and magnetic fields oscillating at right angles to each other, and at right angles to the direction of travel.

In 2014, the media was alight with news that direct evidence of inflation had been discovered. A telescope at the South Pole known as Background Imaging of Cosmic Extragalactic Polarization (BICEP2) had been observing the CMB radiation since 2010. While telescopes in space, such as WMAP and Planck, have provided extremely detailed maps of the *intensity* of this radiation, BICEP2 was looking at something different: the signature of *polarization*.

The polarization of an electromagnetic wave tells us the direction in which the electric field oscillates. Figure 8.2 shows a light wave, comprising oscillating electric and magnetic fields at 90° to each other, and at 90° to the direction of travel. In this case, the polarization is horizontal, because the electric field oscillates in the horizontal direction.

A given light source may be *unpolarized*, meaning that the direction of the electric field changes randomly with time. For example, the light from our Sun is unpolarized. However, when that light strikes the surface of water, the reflection tends to be more horizontally polarized. That is, on average, the electric field wiggles left-and-right more vigorously than up-and-down.

If you are lucky enough to find yourself on a yacht on a sunny day, with lots of bright light glinting off the water, pop on some polaroid sunglasses. These are designed to cut more of the horizontally polarized light, so you see less glare. Tilting your head to the side changes the orientation of the polarizing filter; instead of blocking out the glinting light, the polarizer will let it through.

Why was BICEP2 looking for polarization in the CMB? During inflation, the universe was filled with gravitational waves, shaking space and time themselves. This *gravitational wave background* stretches and squeezes patches of space in particular directions, and this influences the scattering of the radiation that will eventually cool to become today's CMB. In particular, the distortion of space and time leaves a polarization signature. Importantly, many of inflation's early-universe competitors predict that there shouldn't be any gravitational wave background, and so no polarization.

For BICEP2, the stakes were high, as a detection of the signature of polarization would point to gravitational waves coursing through the universe during its earliest epochs, and thereby provide strong evidence for inflation. And, with much fanfare and media hype, BICEP2 announced this detection in March 2014. Inflationary advocates were delighted: over 3 million people on YouTube have watched one of the principle investigators, Assistant Professor Chao-Lin Kuo of Stanford University, dropping by the house of inflation pioneer Andre Linde to give him the good news, accompanied by a celebratory glass of champagne.[6]

However, not everyone was convinced by the BICEP2 result. The discovery had been announced to the media before a scientific paper had been published. To scientists, this is at best an annoyance, and at worst suspicious. We wanted to pore over the details, because this measurement is very subtle – the signal is buried inside a mess of noise from space. How had the BICEP2 team accounted for emission from dust in our Milky Way, which can spoof the inflationary polarization signature?

To account for polarizing dust, BICEP2 needed additional observations of the sky. The Planck satellite had exactly what they needed: a high-resolution map of the dust in the Milky Way, across a range of wavelengths. However, in March 2014, Planck hadn't yet released this data – they had it but were still working on the analysis. So how had BICEP2 accounted for the dust?

As cosmologists probed for more details on this result, the truth came out. The Planck collaboration had shown preliminary versions of the Milky Way maps that BICEP2 needed at a conference, and the BICEP2 team had extracted the data from a PowerPoint slide. When this titbit of information was shared at our weekly astronomy morning tea at The University of Sydney, there was an audible groan from all assembled. It all seemed a bit underhanded, cloak and dagger. And it raised serious doubts: could they extract the data they needed with sufficient accuracy? Had they correctly interpreted the plot? Did the data need further analysis by the Planck team before it could be used?

When the BICEP2 team finally published their result in June 2014, the team had removed all reference to the PowerPoint slide and its Planck data, relying on other methods to estimate the dust. They still reported their measurement, but they noted that the effect of dust would be "better constrained with upcoming data sets".[7]

Meanwhile, other scientists were using the same data sets to do their own analysis. BICEP2 had misinterpreted the Planck data, some claimed, undoing their result. It became clear that BICEP2 had skipped a number of checks and balances, rushing to announce before they got scooped by other cosmologists. In particular, they were worried that the scientists chosen to referee the paper – checking and rechecking their work before it was published – would be exactly the scientists who could scoop them. For the most part, cosmologists were waiting for the Planck collaboration to release their observations.

In a paper in September 2014, and a second paper in early 2015, the Planck collaboration showed that BICEP2's result was toast. As scientist Brian Keating finally admitted, "BICEP2 turned out to be a very precise dust detector."[8] They hadn't seen cosmic polarization. We still don't know whether it exists or not.

A tale of woe, to be sure. But the part that critics of inflation were most concerned about was not the indiscretions of the observers, or even the over-eagerness of inflation's supporters. The real problem was the ease with which, when it was all over, they (ahem) dusted themselves off and carried on. Scientific theories are supposed to predict observations, and to be penalized if those predictions turn out to be wrong. How could they say that the BICEP2 results were strong evidence for inflation, and then say that the retraction of the BICEP2 results was no big deal? This looks like "heads I win, tails we flip again".

For critics, the cat was out of the bag. "The BICEP2 incident . . . revealed a truth about inflationary theory," wrote Paul Steinhardt,[9] a Princeton cosmologist who had helped lay the foundations for inflation in the early 1980s. "The common view is that it is a highly predictive theory . . . [In truth] the inflationary paradigm is so flexible that it is immune to experimental and observational tests."

It is probably correct to say that most cosmologists do not share Steinhardt's scepticism. But neither is Steinhardt a lone voice in the wilderness. This divided opinion is just the way things are at the forefront of science.

If Not Inflation, What?

What else is on the market? In science, all theory *testing* is really theory *comparison*. In trying to test a theory, scientists will look for an alternative model to make a comparison with. If no model can be found, they will often try to make one up. Alternatives to quantum mechanics and Einstein's theory of gravity have been invented by physicists purely for this purpose.

With this in mind, can we imagine an alternative to inflation? What might the early universe have been like? (Of course, many cosmologists have tried to answer these questions. We're just throwing out a few general ideas, rather than surveying the field.)

We seem to be on safe ground in assuming that the very early universe was a sea of tremendously hot matter. On the one hand, quantum mechanics is very much in play, which is often interpreted as painting a random and unruly picture of fundamental physical reality. So, we might expect the early universe to be a messy patchwork of hot and cold, which leads us to the horizon problem.

But remember that physicists also believe that the higher the temperature, the simpler the laws that our universe obeys. Fundamental forces become indistinguishable, and particles interact and transform according to simple rules. Our modern cold universe is much messier in comparison.

So, perhaps conditions at the birth of the universe reflected this simplicity. Perhaps the very early universe's simple laws tend to create the same state in the very early universe, without needing each part of the universe to conspire together to produce the same answer. But that's not how laws of nature usually work: what you get out depends on what you put in. The exception to this principle, where different initial states arrive at the same final state, is usually because the final state is in boring, unchanging equilibrium. But our early universe is changing rapidly, with plenty of free energy that will later be transformed into galaxies and stars and planets.

Moreover, the early universe must not stay *perfectly* smooth into the classical era of cosmology, because it would have remained perfectly smooth for all times. The CMB sky would be perfectly uniform, and there would be no lumps in the universe – like galaxies, stars, planets, and people.

Here, we arrive at an example of a bigger issue. Some parts of our universe show remarkable symmetry, and some don't when they could have. Why? Why is our universe *almost* perfectly uniform when it could have been *perfectly* uniform or much lumpier, filled with black holes from the start? Can we just say, "that's how the universe started", or should we invent a theory like inflation, or is there some other option that we're missing?

On top of this, we've been assuming that space and time are fundamental. That is, space and time aren't, in some way, made of something more basic, more elementary. Some physicists have entertained the idea that space and time are *emergent* phenomena, like water molecules that behave like a fluid when enough of them get together. If quantum ideas apply to space and time themselves, then the fundamental stuff of our universe might be a *spacetime foam* – the geometry of spacetime fluctuates on very small scales. This throws an additional spanner into the works: how can we talk about *initial* conditions if time is just another physical quantity that's just along for the ride, obeying the *real* laws of nature like the rest of us?

And so, dear reader, we must leave you in suspense, the same suspense as the cosmology community. We have tried to push back in time, to say what came before the "hot dense expanding matter" phase of our universe. Inflation, if it happened, seems to solve a number of questions that the standard big bang model leaves open. If it happened. We have ideas, but no standard model of inflation. We have observational hints, but no smoking gun ... and one rather embarrassing misfire. We wait for the next generation of cosmological observations and hope that they are decisive.

9 BALLS FROM LEFT-FIELD

We have covered a lot of ground to get to this point, examining how telescopes see the universe, how scientists analyse observations, and how theoretical physicists use evidence to score points for or against cosmological models. The big bang theory has handled the evidence quite well, while other theories have fallen by the wayside due to one observation or another.

However, an internet search for cosmological models will find many more ideas and hypotheses than can be covered in this book. To try to answer at least some of them, this chapter will give these "also rans" a chance to shine.

Challenging the Standard: The Linear Universe

Our first challenger to the big bang theory is known as the *linear universe*, a universe that expands at a constant rate over its lifetime. Actually, this universe was hotter and denser in the past, so it too is a "big bang" model. But it differs from what we'll call the *standard big bang model*, in ways that we'll describe below.

This is not an idea that we think has a realistic prospect of being an accurate model of the expansion and evolution of the universe, but we want to highlight the fact that it has played the game honourably. As a model, it has been clearly presented by its proponents. They have tried to account for the full range of cosmological data, not just cherry-picking observations that they like. And, most importantly, it has done all this in academic journals.

Before we begin, we need to put our cards on the table: we have skin in this particular game. We have not only followed the details of this particular challenger to the standard big bang model, but have written papers detailing its weaknesses. We're convinced that it doesn't work because we've tried it.

The story begins with a strange coincidence. It's a little technical, but it is the kind of thing that piques the interest of cosmologists.

Within the standard big bang model, the early universe was dominated by the gravitational attraction of matter, but in more recent times (the last billion years or so) the expansion of the universe has changed from *deceleration* to *acceleration*. Today, the universe is not only getting bigger; it's getting bigger at an increasing rate. (We'll discuss why this is happening in Chapter 10.)

Rather strangely, if we average over the deceleration and acceleration phases of the expansion, we get an answer that is very close to zero. That is, the size of the universe today is as if there had been no acceleration or deceleration at all.

This holds true today, but not in the past (deceleration would have dominated) or future (acceleration will dominate). This "zero average acceleration" only occurs in a (cosmically!) short period of time in the entire history of the universe. Why do we happen to live at that time?

We can state the coincidence another way. As we discussed in Chapter 3, a key focus of much of the twentieth century's cosmological effort was measuring the universe's expansion rate, encapsulated in a single number: the Hubble constant. After a few decades in which the uncertainty in this number was a factor of two, modern measurements have arrived at the value of about 70 km/s per Mpc. That's a strange collection of units; here's how to understand it: for every megaparsec (i.e. about 3 million light years) we travel into the universe, the galaxies there are moving away 70 km/s faster.

These units are convenient for astronomers, but there is another way to think about it. One divided by the Hubble constant has units of time, known as the *Hubble time*. This is a rough estimate of the age of the universe, assuming that it has undergone no acceleration or deceleration. Modern measurements of the Hubble constant range from 67 km/s per Mpc to 73 km/s per Mpc, so the Hubble time is somewhere between 13.4 and 14.6 billion years. The weird thing is that's rather close to the actual age of the universe! For the cosmological parameters from the Planck satellite, for example, the difference is about 5%.

Again, if we look at the entire history of the universe in the standard big bang model and ask, "when would the age of the universe be almost equal to the Hubble time?", the answer is a (cosmically) small period of time around about now. Should we be puzzled by this coincidence?

Scientists have a love–hate relationship with coincidences. When faced with a coincidence, we have to ask ourselves a set of questions. Is it just due to chance and our sometimes over-active tendency to see patterns? Is it due to the way we look at the universe, rather than some feature of the universe itself? Or is it really telling us something deep about the world around us? It's hard to know, ahead of time, whether scratching at a particular coincidence will uncover a seam of gold or a pocket of air.

Let's scratch at this particular coincidence. Can we devise a universe in which this coincidence doesn't just hold for a short period of time, but for all times? Then it wouldn't be a coincidence any more. We need to revise the standard big bang model so that the expansion of the universe does not slow down or speed up, but instead is nice and uniform and constant. If the expansion is always constant, then the acceleration of the expansion is always zero, and the age of the universe is always the inverse of the Hubble constant. Problem solved.

This idea has been proposed several times, with various names, such as *linear expansion* or *coasting cosmology*. In its most recent incarnation, it has been labelled the $R_h = ct$ *universe*. The name comes from the fact that the size of the universe (R_h) scales linearly with time (t) at the speed of light (c).

These linear cosmological models inhabit the same mathematical framework as the big bang model, but with a different expansion history. In Einstein's theory of gravity, the acceleration or deceleration of the universe is determined by the push and pull of matter and energy. With so many forms of matter and energy pulling and pushing on the expansion of the universe, how do we get the constant expansion required by the linear universe?

It turns out that we need to fill the universe with yet another form of energy. This strange material pushes whenever the stuff of the universe pulls, and pulls whenever it pushes, ensuring that the net effect on the universe is always zero. This mystery form of energy has not been christened as yet, so we will refer to it as "*dark stuff*".

You might be scratching your head, wondering how a form of energy could have such a bizarre property. But, the linear universe defender may reply, is this any different from so-called "dark matter" and "dark energy"? Actually, yes. As we'll see in Chapter 10, when we look to particle physics for ideas about what dark matter and dark energy *might* be, we get some reasonable candidates. Dark matter might be an as-yet-undiscovered heavy particle, and we know about heavy particles. Dark energy might be *vacuum energy*, and we have a well-tested model for that. But dark stuff, with its ability to know which way the energy and matter in the universe are pulling/pushing, is unlike anything that any particle physicist has ever proposed.

As we have been at pains to point out in this book, the real test of any scientific theory is its ability to make predictions that explain real data. Theoretical physics is just fiction-with-equations until you start making precise, mathematical predictions that we can compare to real data. The $R_h = ct$ model's battle with the big bang is over the following question: is the observed universe better described by linear expansion, or by a period of deceleration followed by acceleration? And the battleground is in the pages of academic journals.

We mentioned that we have "skin in this game". Geraint and his students have published papers showing that "dark stuff" is not just bizarre but unphysical.[1] We have also published a paper that calculates the predictions of early-universe nucleosynthesis in the $R_h = ct$ model, finding that the model does not agree with the abundances we observe in the universe.[2] Others have argued that the model fails to account for observations that show the recent acceleration of the expansion of the universe, and in particular the fact that the Hubble constant has changed over time.[3] As best we can tell, the $R_h = ct$ cosmological model fails to explain the real universe.

So, what do we do with this theory? It may be possible, with a liberal application of ad hoc patching, and some careful tuning, to find a version of the $R_h = ct$ model that explains the data. But this added complexity comes at a price. The (perhaps appealing) simplicity of a linear expansion may be lost among the added bells and whistles of the revisions.

While we disagree with this model, in both its present and past incarnations, this is the way that science should proceed. With publications in journals that focus on theoretical models and predictions of data. That's how you do it.

Dirac Large Number Hypothesis

The advocates of linear models were not the first to wonder whether coincidental numbers tell us about something deep about the cosmos. Paul Dirac noticed some curious numbers of this sort in 1937. He was one of the founders of modern quantum mechanics, extending the work of earlier physicists to include Einstein's special theory of relativity. His theoretical insights led to the prediction of antiparticles, specifically the electron's positively charged twin, the positron. His work on relativistic quantum mechanics earned him the Nobel Prize, shared with Erwin Schrödinger in 1933.

Dirac, following from some earlier insights by Arthur Eddington and Georges Lemaître, was thinking about dimensionless numbers of nature. These are numbers that do not depend on the system of units that we have chosen. For example, the mass of the electron is *not* dimensionless: it is 9.1×10^{-31} kg, 2.0×10^{-30} pounds, or 6.2×10^{-17} femtoslugs; yes, the *slug* is a unit of mass in the British Imperial measurement system, equal to 14.5939 kg. The ratio of the proton mass to the electron mass, on the other hand, is dimensionless: it is 1836.15267, whether we have used kilograms, pounds, slugs, or whatever. For this reason, dimensionless numbers in physics are quite important. They don't depend on our arbitrary choice of units.

When a dimensionless number appears in our equations, and we can calculate its value, it is often a combination of small integers (2, 3, 5, 7, ...) and mathematical constants such as π. For example, suppose we calculate the time it would take for a static uniform ball of matter to collapse under gravity. There will be some relationship between the time to collapse, the gravitational constant, the initial size of the ball, and the mass of the ball. When we use Newton's law of gravity to work out the exact answer, a mathematical constant appears, which in this case happens to be $(\pi\sqrt{2})/4$, which is about 1.11. Because small numbers are being multiplied and divided, these dimensionless numbers tend to be roughly equal to one.

Dirac, however, noted that there were some dimensionless numbers in physics that were not nearly equal to one. They were enormous – billions of billions of billions of times larger than one. For example, consider the forces between protons and electrons. Both of these particles are electrically charged, and so there is an electromagnetic attraction between them. The particles also have mass, and so there is a gravitational force between them. We know that gravity is a much weaker force than electromagnetism, but by how much? Plugging the numbers into the classical formulae for these forces reveals that the electromagnetic attraction between the two particles is stronger than gravity by a factor of around 10^{40}, an enormous dimensionless number.

Dirac also wondered about another ratio. The age of the universe, as best it was known in Dirac's day, is a few billion years. There is also a characteristic time that we find in atomic physics, which represents roughly the amount of time it takes for electrons to shuffle around in atoms. If we ask how many atomic times have passed in the whole history of the universe, we get another enormous dimensionless number: about 10^{39}.

This number, you will note, is not too different from the previous large number. Given the accuracy of the calculation, this struck Dirac as worthy of further investigation. Dirac made a bold proposal: the two numbers are equal. But how?

Dirac's idea is immediately problematic. In our best theories of how gravity and electromagnetism work, the forces between an electron and proton do not change with time. Experiments have looked for evidence that they vary from year to year, but no change has been observed. Similarly, the mass of the electron and the proton don't seem to change. But the age of the universe is always increasing. Like the rest of us, the universe keeps getting older.

This ruins the coincidence. If these two numbers are exactly equal today, then they won't be equal tomorrow, because the universe is a day older but everything else in the equation stays the same. The equation didn't work yesterday, either. A billion years ago, or a billion years into the future, the equation is even more disparate.

Unless. Unless something else in the equation changes over time. If another quantity changes, like the mass of the electron or the speed of light, then we can arrange for the equation to hold at every time. Dirac decided that the most likely culprit was gravity. Newton's gravitational constant, G, Dirac proposed, was not a constant at all but gets smaller as the universe gets older.

And now, to an astrophysicist, the fun really starts. And by fun, we of course mean astrophysics! Making a model of a stable astrophysical object, whether a cluster of galaxies or a neutron star or a planet, involves balancing the various forces on the system. For example, in a star, the crush of gravity is balanced by the outward pressure of the hot gas. In any big object in the universe – basically anything bigger than an asteroid – the gravitational pull of an object on *itself* is crucial to understanding how it holds together. Gravity has a central role in shaping the universe we see around us, from collecting dark matter and gas into galaxies to providing the crush that ignites and contains fiery stellar cores.

So, one does not simply change the gravitational constant and expect the universe to return to business as usual. The change will be written into the observations of the heavens.

Astronomers have looked near and far for the signature of a varying gravitational constant, but nothing has been observed. The spinning of distant pulsars shows no sign of a changing strength of gravity. The formation of the elements in the initial stages of the universe shows no sign of a changing strength of gravity. In fact, in all experiments and observations to date, the fundamental constants behave, well, like constants.

Dirac's idea, alas, is simply not borne out by observations. But the coincidence still remains. Other physicists tried different approaches, including proposing that the gravitational "constant" was really a field, which varies from time to time *and* from place to place. This led to theories of gravity that were different from Einstein's.

One of the most fruitful insights came from Princeton physicist Robert Dicke. He thought carefully about one term in the equation: the age of the universe. But this isn't a fundamental physical quantity. It's just the current time on the cosmic clock, which is always ticking away. Just as we have a location in space – the outskirts of the Milky Way, on the third rock from the Sun, on the big brown continent underneath – we also have a location in time. That's what goes in the equation.

When seen in this light, the *age of the universe* is a rather parochial quantity. It's the time that *humankind* sees on the cosmic clock. Why would the fundamental constants of nature care about humanity's curtain call on the cosmic stage?

Dicke reasoned as follows: Humanity didn't just turn up in the universe at some random time. We couldn't have existed when the clock said a million years, because the universe was pure hydrogen and helium – no stars, no planets, no solid objects of any kind, in fact.

So, what would a life form expect to see on their cosmic clock? They would probably have to wait until at least the first generation of main sequence stars had made the elements from which life is made: carbon, oxygen, phosphorus, sulphur, and so on. Dicke took a simple model of how stars work, used it to relate the lifetime of a star to the fundamental constants of nature, and *voilà*! The lifetime of a star, in atomic times, is roughly equal to the ratio of the electric to gravitational forces for protons and electrons. Dirac's coincidence is explained.

Well, almost. The calculation gives a rough idea of the *minimum* amount of time it would take for the universe to make the elements for life. But this isn't the same as when life might expect to exist. If life spreads throughout the galaxy for the next trillion years, then most life forms won't see Dirac's coincidence.

Dicke's explanation raises a complication for any aspiring fundamental-constant-coincidence-hunter. A given coincidence might point towards deeper, more amazing physics, or it might be a consequence of our existence as life forms.

To work out which is which, physicists in the 1970s began to investigate the connections between life and the fundamental constants of nature. As we have discussed in our previous book, *A Fortunate Universe: Life in a Finely Tuned Cosmos*, physicists have found that life relies on a fascinating set of relationships between these numbers. Some very small changes can result in a universe without structure, without atoms, without stars, and/ or without some other feature upon which life as we know it, or life as we can imagine it, depends.

For our purposes, and for most physicists today, Dirac's idea is dead. The coincidence is not a clue to changing constants of nature; the best explanation is in terms of the requirements of life. But it did lead to some very interesting ideas and calculations. There's a lesson here: bold ideas can be fruitful even if they are false.

Redshift Quantization

In Chapter 3, we discussed the redshift of light from distant galaxies, and how this has played a central role in modern cosmology. We discussed the claims of some astronomers that the redshifts of *quasars* – very compact, bright sources of light in the distant universe – are not due to the expansion of the universe but are due to some intrinsic physical process. Such ideas, even if correct, don't affect the case for the big bang, which relies on the redshifts of *galaxies*.

Critics of the big bang theory have raised another problem with redshifts: they claim that they are *quantized*. That is, when we measure redshifts of a whole bunch of galaxies, we don't find a random assortment of values. There is a tendency for redshifts to have values that are evenly spaced or *periodic*. If the universe is randomly scattered with galaxies, on average, and redshift is a measure of distance, shouldn't the redshifts of galaxies also be random?

In the 1960s and 1970s, K. G. Karlsson and Geoffrey Burbidge used a survey of 166 newly discovered quasars to claim that, rather than being random, their redshifts could be described by a simple formula: they prefer the values 0.06, 0.30, 0.60, 0.96, 1.41, 1.96, ..., with the values evenly spaced on a logarithmic scale.[4] Readers with a weak mathematical constitution should look away now; the formula is $z_{quasars} = 1.06 \times 10^{0.089n} - 1$, where n is a whole number (0, 1, 2, 3, ...).

Is this a problem for the big bang? Not at all. We can repeat our comments from the end of Chapter 3. Even if these claims were correct, this would only show that quasars don't obey Hubble's Law. It would not show that Hubble's Law is false, because Hubble's Law is about galaxies. We would have to rewrite the textbook about the astrophysics of quasars, but Hubble's Law, and so the big bang theory, would remain untouched.

Regardless, we can check Karlsson's formula using vastly larger modern surveys of quasars, including the 2-degree Field Quasar Redshift Survey (2dFQRS) and Sloan Digital Sky Survey (SDSS). In 2010, Repin, Komberg, and Lukash used redshift data from 85,000 quasars to show that there was no evidence for special, preferred redshifts.[5]

A more interesting claim, for our purposes, was made first by William Tifft, an astronomer from the University of Arizona.[6] Working in the 1970s, Tifft and others painstakingly measured the redshifts of individual galaxies, one at a time, building up a catalogue. They claimed that galaxies exhibit a preference for certain redshifts.

Here, we must distinguish carefully between two separate claims that have been made about galaxy redshifts.

Claim 1: Pairs, Groups and Clusters of Galaxies Show Preferred Redshifts, Relative to Each Other

This was the original quantized redshift claim. Studies of galaxy clusters seemed to show that, if you were sitting in the middle of the cluster, the other galaxies would prefer to be moving away from you at some multiple of 72 km/s, that is, 72 km/s, 144 km/s, 216 km/s, 288 km/s, 360 km/s, ...

Further evidence came from studies of pairs of galaxies. Once again, put yourself on a galaxy, and look at a nearby galaxy, and (supposedly) you're likely to find a preference for multiples of 72 km/s again. Tifft and others were surprised to see galaxy pairs clumped into distinct, quantized redshifts. This is seemingly at odds with the smooth, homogeneous universe of the big bang.[7]

Does this show that galaxy redshifts can't be due to the expansion of space? Not really. Here's why.

It will be convenient to teach you the units of distance that astronomers actually use. A *parsec* is defined with relation to the technique of parallax, discussed in Chapter 3. As a reminder, one parsec is equal to about 3.26 light years, or about 30 trillion kilometres.

Now, that's a long way, and much larger than any human imagination can grasp. Here's a beginner's guide to using parsecs.

- The distance to the Sun is about a 20,000th of a parsec.
- The distance to the nearest stars is a couple of parsecs.
- The distance to the centre of the Milky Way is about 10 kiloparsecs (kpc). That is, galaxies are typically tens of thousands of parsecs across.
- The distance to the nearest large galaxy (Andromeda) is about 1 million parsecs, or a megaparsec (Mpc).
- The distance to the furthest galaxies we can see is tens of billions of parsecs, that is, tens of gigaparsecs (Gpc).

With this very brief introduction, we can ask: in our expanding universe, how far away is a galaxy that is moving away from us at 72 km/s? Given Hubble's Law, we can calculate that the distance is about one megaparsec (Mpc). That is, it is about the distance to the nearest galaxy to the Milky Way.

Is the universe supposed to be smooth and homogeneous on this scale? No! Imaging grabbing a big box that is 1 Mpc on the side, and scooping up a random bit of the universe. Sometimes, the box will contain a galaxy or two, and so have much *more* matter than average. Sometimes it won't have any galaxies, and so will have much *less* matter than average. The box is too small to scoop a representative sample.

The existence of structure on the scale of a few megaparsecs is not at all a problem for the big bang. Galaxies don't make their own matter – they must collect matter from their surroundings. How large a region must be emptied out in order to make a typical galaxy today? About a 1 Mpc across. We can hardly be surprised if this is a typical distance between galaxies today.

Moreover, Tifft *doesn't* deny Hubble's Law; in fact, he uses it to identify pairs of galaxies. He assumes that if two galaxies are close to each other on the night sky, *and* have similar redshifts, then they are close to each other in space. He doesn't consider redshift quantization to overthrow the big bang, but rather to show that something is missing from our understanding of how galaxies form.

And even then, the data is not particularly convincing. Tifft undertook his observations in the 1970s. Galaxy survey technology and techniques rapidly evolved in the 1980s and 1990s, and astronomers were able to obtain more accurate redshifts for larger samples of galaxies. A study by Nordgren, Terzian, and Salpeter in 1996 failed to find evidence for small-scale redshift quantization.[8] Few astronomers were convinced by the first set of observations in the 1970s, and so this issue has received little attention.

Claim 2: Galaxies from Across the Universe Show a Preference for Certain Redshifts, on Large Scales (>100 Mpc)

As galaxy surveys grew in size, a similar-sounding but much more important claim was made. In 1990, Thomas Broadhurst and colleagues reported in the prestigious journal *Nature* that galaxy redshifts were far from random on properly cosmological scales.[9] By combining four large (for the time) and deep surveys, with a total of 396 galaxies, they demonstrated a striking preference for galaxies to be found at distances that are multiples of 180 Mpc.

Now, that's much bigger than the distance to the nearest galaxy. It's bigger than the biggest structure we've discovered in the universe, which is the Laniakea Supercluster, a 77 Mpc-wide collection of around 100,000 galaxies, including our Milky Way.

This result has attracted plenty of attention from the cosmology community. Broadhurst's paper has been cited by almost 600 subsequent scientific publications. The result was reinforced in 1997 by the work of Jaan Einasto and colleagues,[10] who used a survey of 869 galaxy clusters to reveal structure in the universe on the same scale as that found by Broadhurst.

This claim has bite. The data is better, the periodicity is more obvious, and this phenomenon strikes at the foundations of the big bang. Does it show that the universe is not homogeneous?

A number of points need to be made. These matters are still under discussion by cosmologists. Opinions are not settled.

The correct use of statistics is always a concern, especially when data sets are relatively small. For example, the eminent cosmologists Nick Kaiser and John Peacock argued in 1991 that Broadhurst's result could be explained by random alignment of a small number of galaxy clusters.[11] However, computer simulations by Yoshida and collaborators in 2001 concluded that Broadhurst's structures were very unlikely in the standard big bang model of our universe.[12] Unfortunately, the work of Yoshida has not been followed up with larger and/or more detailed modern simulations.

Further, the structures of Broadhurst have not unambiguously turned up in larger galaxy surveys. The 229,193 galaxies of the 2dF survey, for example, don't show the dramatic picket-fence-like structure of earlier, smaller surveys. More recent claims about the galaxy population have changed in light of this data: rather than saying that the redshifts are quantized, proponents claim that, as well as the usual uniform distribution of galaxies, there is a slight periodic signal on top. But the evidence for this is not particularly convincing.

If we're looking for a slight periodic effect, then there is a plausible candidate within the standard big bang theory. Given the history of the universe, we can ask: what's the largest scale in the universe that could feel the squeeze of gravity? This is given the technical name of *baryon acoustic oscillations*, and in recent years the signature of this effect has been found in surveys of the universe. In fact, five galaxy surveys, measuring hundreds of thousands of galaxies, showed that there is a small enhancement of structure in the universe on scales of 150 Mpc. Now, no one thinks that Broadhurst actually measured baryon acoustic oscillations. But this does show that one must be careful with deriving predictions of smoothness and structure from the big bang theory.

One of the problems with attempts to use redshift quantization against the big bang is that the data is often compared to a *random* sample. But the big bang theory does not predict that galaxies will be randomly scattered across the universe on small scales. Simulations of how galaxies form show the *cosmic web*: beginning with the almost smooth early universe, gravity causes matter to collapse into sheets, filaments, and dense knots, as shown in Figure 6.6 in Chapter 6. The distribution is far from random.

In summary, let's try to bring some order to this meandering section.

- In an expanding universe, light from a galaxy or quasar could be stretched by any amount. That is, it could have any redshift.
- Some astronomers have claimed that galaxies or quasars show a particular fondness for certain redshifts, especially evenly spaced (periodic) redshifts. This seems to contradict a tenet of the big bang – the universe is smooth and homogeneous on large scales.
- Claims about quasars don't affect the *evidence* for the big bang theory, as we explained at the end of Chapter 3.
- Claims about the redshifts of galaxies that are a few megaparsecs away are irrelevant as well; the universe is obviously not smooth on those scales.

- The most important claim is that galaxy redshifts are quantized on scales of hundreds of megaparsecs. But recent, larger galaxy surveys have not supported claims made in the early 1990s. More recent reports of quantization are more restrained – there is a small periodic *component* to the distribution of galaxy redshifts. But this is less convincing, and also not obviously inconsistent with the structure that forms in the aftermath of the big bang.

White Hole Cosmology

One popular misconception of the big bang model is that it describes an explosion. Supposedly, the big bang sent matter flying through empty space away from the site of the explosion. When someone asks, "where did the big bang happen?", they probably have this misunderstanding in mind.

Nevertheless, given that galaxies are moving away from us (as we saw in Chapter 3), some scientists have tried to produce an alternative model of our universe as some kind of explosion. One particularly interesting example is the work of physicists Joel Smoller and Blake Temple in 2003.[13]

They construct the following marvellous scenario. For some reason, the universe is born as an expanding explosion inside a black hole. At the leading edge of the expansion is a shock wave, which propagates into the empty space around the original black hole. In its wake, the universe – strange, but true – looks essentially identical to the aftermath of the big bang. The main difference between the two models is that, somewhere out there, there is a shock wave that marks the boundary between matter and empty space.

Could we live in that universe? While the model is ingeniously constructed within Einstein's theory of relativity, the problem is that we see no sign whatsoever of this shock wave. The model must suppose that the initial size of the shock wave was large enough that it is currently too far away for us to see. Smoller and Temple note that there are a few parameters that we need to completely specify the properties of the shock wave.

Unfortunately, these numbers are completely unobservable since we can't see the shock. They call their equations a "rough qualitative model" for this reason.

This makes the model, from an observational standpoint, an unnecessary complication of the big bang theory. While this is an interesting scenario, in one sense it is not surprising. We only ever see a finite piece of the universe because light can only travel so far since the beginning. We can't know what lies beyond the edges of the *observable universe*. So, of course, we can put an expanding shock wave out there. But we could also put a million monkeys on bicycles out there, and our telescopes would never know. The 13.8-billion-year history of the observable universe unfolds just the same.

A similar-sounding model was proposed in the early 1990s by physicist Russell Humphreys. Humphreys is a young-Earth creationist: he believes that the first few chapters of the book of Genesis teach that the universe was created around 6,000 years ago.[14] For an astronomer, this immediately creates a problem. As we explained in Chapter 3, we can see stars and galaxies that are more than 6,000 light years away. On a clear night you can see the central bulge of our own Milky Way galaxy, 24,000 light years away. With a reasonable backyard telescope, you can see galaxies that are millions of light years away. But that means that the light we see departed the star or galaxy millions of years ago, doesn't it?

Here is Humphreys's idea. In Einstein's general theory of relativity, the warping of space and time can cause clocks to run at different rates. Imagine a pair of astronaut twins who are observing a massive black hole, far enough away to be safe from the intense gravitational forces. One sets off on a journey to get nearer to the black hole, while the other remains where they are. The returning traveller would find that they are younger than the homebody. This was used to great (and fairly plausible) effect in the blockbuster movie *Interstellar*.

In Humphreys's universe, the universe expands out of a *white hole*, which is essentially a black hole running in reverse. It spits out matter instead of pulling it in. However, unlike the shock wave model of Smoller and Temple, the familiar post-big bang universe is on the *outside* of the white hole. Rather than being too far away to see, the white hole is in our past: it spat out all its matter and evaporated. One of the last objects to be spat out by the white hole was the Earth. Thanks to the extreme slowing of clocks, billions of years pass in the distant universe, while time on Earth effectively stands still. Clocks on Earth have only ticked off 6,000 years, but in this time, light from distant galaxies has had billions of years to make the journey to us.

Now, it would be easy to simply dismiss this model because of its religious motivation. But this is *at most* a reason to be suspicious. It might explain why Humphreys is wrong, but it doesn't explain why his *idea* is wrong. Wrong scientific ideas aren't wrong *because* the scientist was biased. They're wrong because the universe isn't like that.

Let's press this point a little. We don't test scientific ideas by interrogating their proposers; we test scientific ideas by interrogating the universe. After all, scientific ideas often arise in a spark of creativity that leaves the scientist without much idea of where they came from. The great quantum physicist Paul Dirac once said:

> Many of my papers were consequences of an idea that had come to me rather accidentally ... which just came out of the blue. I could not very well say how it occurred to me. And I felt that work of this kind was a rather undeserved success.[15]

Scientific ideas have come from unusual circumstances. In 1862, August Kekulé first proposed that the molecule benzene contained carbon atoms joined in a ring, an idea that came to him in a dream about a snake swallowing its tail. We're reminded of the story of The Beatles' song *Yesterday*: Paul McCartney dreamed the melody, and when he woke up had to ask his fellow Beatles whether he had really composed it, or just unconsciously remembered it. Kip Thorne's research into wormholes in general relativity was motivated by Carl Sagan, who wanted to put them in his science fiction story *Contact*.

Psychologically analysing scientists is not a good way of uncovering the secrets of the universe. The philosopher Antony Flew called this the *subject/motive shift*; C. S. Lewis called it *Bulverism*:

> ... *you must show that a man is wrong before you start explaining why he is wrong. The modern method is to assume without discussion that he is wrong and then distract his attention from this (the only real issue) by busily explaining how he became so silly. I call it Bulverism. ... Bulverism is a truly democratic game in the sense that all can play it all day long, and that it gives no unfair privilege to the small and offensive minority who reason. ...*
>
> *I see Bulverism at work in every political argument. The capitalists must be bad economists because we know why they want capitalism, and equally the Communists must be bad economists because we know why they want Communism. Thus, the Bulverists on both sides. In reality, of course, either the doctrines of the capitalists are false, or the doctrines of the Communists, or both; but you can only find out the rights and wrongs by reasoning – never by being rude about your opponent's psychology.*[16]

As we have pointed out, if someone wants to tell us how the natural universe is, then we ask just two things. Firstly, that their scientific theory is about natural things and their effects. In short, is this a way that the physical world could be? And, despite appearances, Humphreys's model invokes only natural laws and natural things. Remember that *all* scientific models must simply state their initial conditions, so Humphreys's belief that God is the cause of the initial conditions of his model does not make the model supernatural. You could believe that about *any* cosmological model, but that doesn't make all cosmological models religious. And, secondly, we ask that their scientific idea is precise enough that we can ask questions such as, "If I performed this experiment or made that observation, what would I expect to see?"

We can do this with Humphreys's model. In the years after its proposal, it was criticized on a number of fronts. Most of the theoretical focus was on the claim that there would be a stage in which clocks on Earth stood still while distant clocks ticked away. Did this really happen in Humphreys's equations? Several physicists, including the eminent cosmologist Don Page, said no.[17] Humphreys's model is just a confusing rewrite of the big bang, in which there is no such physical time dilation. The effects that Humphreys claimed were the result of a mathematical error in the way he extracted predictions from his equations. Humphreys disagreed.[18] Back and forth it went.

This disagreement is too technical to explain here; for the record, we're with Don Page on this one. But we can explain another problem with the model, one pointed out by fellow young-Earth creationist and physicist John Hartnett.[19]

Consider the Andromeda galaxy, which is 2.5 million light years away from Earth. (Humphreys doesn't dispute these distances.) Supposedly, while light made this long journey, a very short time (about a day) passed on Earth. This requires an enormous amount of gravitational slowing of time, which in turn implies extreme warping of space and time. But, as Hartnett points out, there is no evidence of the aftermath of this phenomenal warping. For example, the light from Andromeda itself arrives at us essentially unchanged, with no sign of redshift or blueshift from intense gravitational fields. Hartnett says, "I don't believe the relative clock rates can be tinkered with to achieve a sensible result."

While even fellow young-Earth creationists seem to have abandoned Humphreys's idea, others have proposed other methods for light from the distant universe to arrive at Earth rather quickly. Our purpose here is not to discuss every such idea. Our point is that, so long as you have an idea about the universe that is precise enough to predict observations, we can think about and test your idea.

But what about human prejudices and biases? Don't they affect people? Of course. The interesting thing is *how* they affect a scientific theory. If a theory about the universe is based on an incorrect assumption, then so long as this is made precise and testable, we can hope that there will be a conflict with the actual universe. It doesn't always happen – the prediction might not be practically testable. But if it does happen, watch carefully how the defender of the theory reacts.

They may insist that their idea really does predict the data, or that the data is somehow mistaken. This is not uncommon. It just means we have to pull our socks up and double check everything. But if things look really dire, they may break the rules. They may take refuge in vagueness, refusing to give enough detail about their idea to allow us to test it. Or they may introduce new, ad hoc assumptions designed only to save the theory from facing the real universe. This is often accompanied by accusations of mainstream conspiracy and oppression.

When this happens, we aren't doing science any more. There is a certain dignity in a failed scientific theory: it played the game, it faced the data, it made precise predictions, and it accepted its defeat with grace. At least it didn't cheat. It didn't retreat into vagueness, like an astrologer predicting that this week you'll have dealings with money and experience emotions. And it didn't keep adding new "principles", like a child who keeps changing the rules of the game so that they always win. Failed scientific theories, we salute you!

Cosmic Microwave Background Scepticism

In 2002, a Professor of Radiology from Ohio called Pierre-Marie Robitaille paid for a full-page ad in the *New York Times*, at the cost of almost a year's salary. The reason, he explained, was "I am at a loss in dealing with the scientific publication of this material." The ad claimed that the Sun, rather than being gaseous, was in fact a liquid. Further, it called into question the cosmic origin of the cosmic microwave background, claiming "it will eventually be discovered that the signal of the COBE satellite is not associated with the 'temperature of the universe,' but is produced by the oceans."

In the years that followed, he published papers with titles such as "COBE: A Radiological Analysis" and "WMAP: A Radiological Analysis" in a journal known for presenting

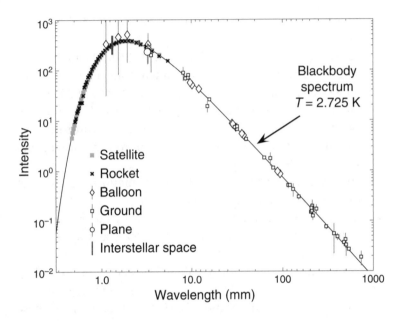

Figure 9.1. Measurements of the cosmic microwave background spectrum by many different experiments. The underlying solid line is the Planck blackbody spectrum at a temperature of 2.725 K.

non-mainstream ideas. In these papers, he applies his medical expertise to astronomical instrumentation. He also produces videos of his ideas for YouTube; the comments section is very enthusiastic.

Whilst published, his ideas have had no impact on mainstream cosmology, with his papers being completely uncited by other researchers. And it's not difficult to see why. The idea that the various instruments that have tried to measure the CMB have in fact measured the ocean is, to say the least, far-fetched.

Firstly, many independent instruments have measured the spectrum of the CMB. Figure 9.1 shows the result of 147 measurements of the CMB spectrum from dozens of instruments at a variety of locations:

- the COBE satellite, 900 km above Earth. (The same as Figure 5.2, but without the exaggerated error bars. Some of these grey squares are hidden behind black triangles.);

- a Canadian rocket that reached an altitude of 250 km over New Mexico (COBRA);
- 15 measurements from high-altitude (>35 km) balloons;
- 37 ground-based measurements taken from such places as Italy's *Campo Imperatore* alpine meadow (altitude 2,000 m, 50 km from the ocean), the Amundsen–Scott South Pole station (altitude 2,835 m, 1,000 km from the ocean), California's Mount White (altitude 3,800 m, 300 km from the ocean), and small-town New Jersey, USA;
- one measurement from an instrument mounted in a Learjet that flew to an altitude of 15 km;[20] and
- the best modern results from measurements of cyanogen on sightlines to 16 separate stars (recall McKellar's work in Chapter 5).[21]

These instruments were independently designed and tested by dozens of professional physicists and engineers over several decades. Robitaille's accusation is that, not only were these instruments inadequately shielded and so accidentally measured the ocean, but also that they all measured *exactly* the same spectral properties from the ocean. Somehow, the ocean – whose temperature varies between 270 and 300 K – can perfectly mimic the spectral properties of a 2.725 K blackbody, over a large range (a factor of a thousand!) in wavelength, whether your instrument is on the South Pole, at sea level, in a balloon, in space, or pointing out of the modified safety hatch of a jet. And, further, regardless of your efforts to shield external radiation.

Secondly, we know what the absorption and emission properties of the ocean are. We have instruments that can measure light at a variety of wavelengths. We've pointed them at the ocean. It doesn't look like Figure 9.1. By contrast, no model, calculation or measurement showing *how* the ocean produces CMB data has been forthcoming from Robitaille.

Finally, none of this would dissuade Robitaille, who claims that everyone else is doing thermodynamics wrong. Physicists incorrectly assume that liquids and gases can produce blackbody radiation, he says. While it's one thing to question cosmology, thermodynamics has been a foundation of physics, and numerous branches of engineering, for over 200 years. Where Robitaille should be rewriting thermodynamics text-books and performing experiments, he instead publishes papers like "Blackbody Radiation in Optically Thick Gases?". This paper, while standing in opposition to over a century of thermodynamical experiment, theory, and practice, contains no equations, no measurements, no data, and has made no impact on physics.

A Never-Ending Stream of Ideas

One of the joys of being a professional cosmologist is the oppor-tunity to share evidence and ideas with the public, especially through public talks. At the end of the talk, we open the floor to questions from the audience, which often demonstrate the fas-cination of the public with these topics. But occasionally some-one in the audience will stand up, clear their throat, and hold forth on their disdain for the big bang. Protesting in the name of logic and common sense, they put forward their ideas of how the universe works. (These questions are tailor-made for the response, "I'll take that as a comment.")

And then there are the letters, emails, and even phone calls out of the blue. (We're not quite sure why, but retired engineers are significantly over-represented among amateur cosmologists. Not that there's anything wrong with that!) This correspondence resembles the comments made at the end of our public talks, but often elaborated with diagrams and doodles, and tales of oscil-lating this and irrefutable that.

Usually missing is the one thing that any scientific revolution requires: clear mathematical structure, leading to precise pre-dictions, able to be tested. The words and the pictures come with a request: "I know my idea is right; I just need some help adding some mathematics." Scientists usually file such correspondence in a special cylindrical storage device at the end of our desks, emptied daily. (Actually, the university brains trust decided that academics are no longer allowed to have bins in their office. It is more cost-effective to have communal bins in corridors and get senior professors to empty their own bins. Someone gets paid to make decisions like that.)

If you want your revolutionary cosmological ideas to be read and understood by those in academia, then don't take this approach. Make some predictions and get them into the academic press!

10 HOW TO DO BETTER THAN THE BIG BANG

We've come a long way, winding through the observations that support the big bang theory, and highlighting where rivals have fallen short. We've looked at the darkness of the night sky, the redshift of galaxies, the slowing of supernovae, the dimming of galaxies, the cosmic microwave background, the Lyman alpha forest, the cosmic elemental abundances, and a few mysteries that led us towards the theory of cosmic inflation.

But we don't want to leave you with the impression that cosmology is all sewn up, with no challenges still to address and no questions yet to answer. If you have come this far, and your own cosmological ideas have cleared the observational hurdles we have laid before it, adequately explaining the properties of the universe without resorting to ad hoc assumptions, then this is the place you really enter into battle. You have the opportunity to slay the big bang theory and put your model on the throne.

This chapter will look at some loose ends, untested predictions, observational puzzles, and missing pieces. We'll work our way backwards in time, towards the biggest puzzle of them all: the beginning.

The Dark Universe

We mentioned in Chapter 1 that modern cosmology has made some rather bizarre claims in the past few decades. Perhaps the most alarming is that 95% of the energy in the universe is in an unknown form. Ordinary matter, that is, the matter that makes up atoms, accounts for only 5% of the energy budget of the universe. Some of our observations of the universe require two additional – and unusual – cosmic constituents, known as *dark matter* and *dark energy*.

This is unsettling. No cosmologist is happy with this situation. We find it amusing when critics of the big bang point to the dark sector as if it were some kind of conspiracy, because cosmologists would dump the dark components in a moment. If someone convinced us that the universe could be explained convincingly – without raising even bigger problems – using only ordinary matter, we would gleefully throw all our dark matter and dark energy models in the garbage bin. At the moment, for most cosmologists (though not all, as we'll see), dark matter and dark energy are part of our best theory of the universe only because the alternative models are even worse.

You might think that it's awfully late in the book to bring up an embarrassment of this magnitude. Have we been hiding it away, keeping it under wraps, presenting the parts of the big bang theory that we like and hoping you didn't read this far? Well, here is the remarkable thing: *it hasn't come up*. The big bang's predictions discussed thus far do *not* depend on whether there is dark matter and dark energy in the universe or not. Its explanation of the darkness of the night sky (Chapter 2), Hubble's Law (Chapter 3), supernovae slowing and galaxy dimming (Chapter 4), the cosmic microwave background blackbody spectrum and temperature–redshift relation (Chapter 5), the Lyman alpha forest's map of neutral hydrogen (Chapter 6), and the creation of elements in the big bang (Chapter 7) do *not* depend on the amount of dark matter and dark energy.

Well, almost. Firstly, for the Lyman alpha forest, the existence of the forest doesn't depend on the dark sector, though the details of the pattern depend on the formation of galaxies, which we discuss below. Secondly, Figure 3.8 in Chapter 3 shows Hubble's Law, measured out to very large distances using supernovae explosions. As you can see in the plot, the trend is not perfectly straight. The *curve* in this line at large distances depends on the rate of expansion in the universe's past, and so depends on the universe's energy content. Stay tuned for more on that soon.

But apart from that, the predictions of the big bang model discussed thus far would remain completely untouched if we discovered tomorrow that dark matter and dark energy don't exist. The basic big bang picture – in the past, the universe was hotter and denser – is enough to explain these important facts about our universe.

Does this mean that dark matter and dark energy aren't really an embarrassment? Not at all! Scientific theories, as we constantly stress, can't be happy to explain *some* of the data and ignore the rest. If we have observed ten facts about the universe, and your model explains only nine, then *something* is wrong. The hard question now is: is your model *close*? Is there a *small* change that will naturally and neatly explain that tenth fact? Or are such small changes ad hoc and contrived? Is that tenth fact telling us to throw everything away, and start again from scratch?

So, what were the observations that caused cosmologists to appeal to the dark sector? In particular, how can we make pronouncements about dark matter and dark energy without knowing what they actually are? How is this different than just making things up?

Einstein's general theory of relativity describes the evolution of matter in the universe. In principle, your task is straightforward: to calculate how spacetime and its contents change, just tell the equations where the matter and energy are, turn the big handle, and see what gets spat out the other side. (In practice, brace yourself for a big barrel of mathematics.)

We should clarify some cosmologist terminology at this point. Gravity feels the effect of all forms of energy. When we talk about *matter*, whether ordinary matter or dark matter, we mean a substance that has the vast majority of its energy locked up in the form of *mass*. Gravity doesn't care about what kind of mass it is. If that seems a little strange, imagine a cardboard box sitting on a weighing scale, sealed tightly so you cannot see inside. The reading on the scale is 10 kg. Now, does the box contain 10 kg of lead, or 10 kg of gold? You can't tell. The two materials have an identical *gravitational* effect on the scale. You would need to prise open the lid of the box and shine a torch inside, or test the density of the material, or take an individual atom and peek at how many protons and neutrons are in the nucleus.

Gravity is blind to the details of the kind of matter, only "feeling" the bulk quantity of the total mass. Applying this to the universe, observing the effect of gravity on the motions of stars and galaxies, or the deflection of light in a gravitational lens, can tell us how much matter is present, but not what kind of matter it is.

When astronomers try to measure the amount of matter in galaxies and clusters of galaxies, we have two methods. We can observe the consequences of gravity, such as the speed of stars in their orbits. Or, we can infer the amount of matter from starlight and observations of interstellar gas. What we find, for every galaxy and cluster in the universe, is a shortfall – there is more matter than the ordinary matter inferred from stars and gas. Searches for ordinary matter in a dark form, such as small planets or even small black holes, have been unsuccessful. As far as we can tell, only 16% of the matter in the universe is ordinary, protons–neutrons–electrons matter; the rest is dark.

Even more convincing evidence comes from the pattern of hot and cold patches in the cosmic microwave background. In these patches, we are seeing under- and over-dense regions in the 380,000-year-old universe. These regions have been pulled by gravity and pushed by the pressure of the hot gas and photons. So, from the pattern, we can infer the amount of ordinary matter (which experiences both pull and push) and dark matter (which experiences only the pull of gravity). We find an answer consistent with the answer from galaxies and clusters: 16% ordinary matter, and the rest is dark.

And then there is dark energy. Unlike matter, however, dark energy isn't as simple as *energy = mass*. We can think of dark energy as a *field* that is spread throughout space. The curious feature of this field is that, as the universe expands, the amount of energy in any given cubic metre of space remains constant. That is, the energy does not dilute as space expands. A useful way to think about dark energy, and one that may be close to the truth, is that dark energy is an amount of energy associated with empty space. This is equivalent to the cosmological constant, proposed by Einstein in 1917 to make a non-expanding, static universe. When we feed dark energy into Einstein's equations, we find a curious effect. Unlike matter, which causes the expansion of the universe to decelerate, dark energy causes the expansion to accelerate.

As mentioned above, Figure 3.8 in Chapter 3 shows Hubble's Law, measured out to very large distances using supernovae explosions, and the line is not perfectly straight. When cosmologists used this curve to infer the contents of the universe, they found that they had to add a significant quantity of dark energy for the model to adequately fit the data. Other observational probes have reached the same conclusion: about 70% of the universe is dark energy, 25% is dark matter, and only 5% is ordinary matter.

What are they? In the dark sector, with just a few clues from astronomical observation and no experimental clues from particle physics, speculation and guesswork abound. We have too many ideas and not enough data to help us decide between them. But our chances of uncovering the identity of dark matter are more promising, so we will start there.

We have met some of the fundamental pieces of our universe already: quarks, electrons, photons, etc. While the standard model of particle physics is among the most successful and accurate physical theories we have, it cannot be complete. There are open questions: why do neutrinos have mass, why are there three families of quarks and leptons, and how does gravity fit into this picture?

Physicists have looked for ways to extend the standard model, in particular looking for ways to make the equations more mathematically symmetric. One much-discussed idea is known as *supersymmetry*, which proposes that there is a cousin for every existing particle: a squark for a quark, a selectron for an electron, a photino for a photon, and so on. While we haven't seen any evidence for these particles in experiments yet, it does provide a neat model for dark matter. If one of these supersymmetric cousins has a sufficiently large mass, is produced in sufficient numbers in the very early universe, and remains stable and unaffected by other matter until today, then it will behave just like the dark matter we require. In fact, many of the proposed extensions to the standard model have nominated candidates for dark matter with exciting names such as axions, hooperons, sterile neutrinos, WIMPs, and Q-balls. If you are going to propose a new dark matter candidate, make sure you give it an exciting name!

Could we take (dark) matters into our own hands and make one of these particles ourselves? In fact, that is just what we hope the Large Hadron Collider (LHC) will do. The LHC smashes together energetic protons and antiprotons, and particle physicists wait with bated breath. The collisions create a spray of particles inside a detector, and we check for evidence of particles that aren't made of the standard model ingredients. So far, nature has refused to play ball. But the hope is that, as the LHC creates more collisions at higher and higher energy, eventually the LHC will become a dark matter factory. Once we can make these particles, we should be able to measure their properties and see if any of our existing dark matter theories are correct. As of yet, physicists sit and wait. In fact, with respect to some dark matter candidates – particularly supersymmetric ones – the LHC's silence is becoming rather awkward.

We should note that not all cosmologists think that dark matter is a good idea. The supposed evidence for dark matter, the motions of stars and galaxies, the gravitational lensing of light from distant sources, and the cosmic microwave background, is actually evidence that we have gotten gravity and motion wrong.

This idea is known as *modified-Newtonian dynamics* (MOND), and in its simplest form it states that Newton wasn't quite right when he claimed that the acceleration of a particle is proportional to the force applied. When accelerations are very small, the relationship changes: force is proportional to the acceleration *squared*. This simple idea, it turns out, does a reasonable job in explaining why we think that we are missing so much of the mass in galaxies.

The key appeal of MOND is that we only need ordinary matter, the matter we can actually see, to explain the universe. But opponents are not happy with the ad-hocness of aspects of the theory, the messiness of the underlying mathematics, and the tweaking of parameters required to make MOND work. Most cosmologists would bet on dark matter, but MOND advocates show little sign of slowing down. There has been no shortage of scrutiny; in fact, dark matter advocates have at times been a bit too eager to find the silver bullet that disproves MOND once and for all, while ignoring dark matter's less than perfect track record. Nevertheless – take note, cosmic revolutionary – MOND advocates are going about their business in the scientific way: formulating their ideas using equations, using those equations to explain data, and publishing in academic journals. If MOND goes the way of the dodo, it will be an honourable end.

The dark *energy* predicament is worse. In the form of the *cosmological constant,* dark energy was postulated and then unceremoniously dumped by Einstein in the 1920s, doomed to lay unwanted and unloved in the equations of cosmology for around 70 years. The amount of dark energy required to explain cosmological observations appeared to be zero.

But all of this changed in the 1990s when the distant cosmos was unveiled by bigger and better telescopes. As telescopes swept the sky for more and more supernovae, we could extend Hubble's Law into the very distant universe, as shown in Figure 3.8. We reached back into the history of the expansion of space, which depends on what the universe is made of. And cosmologists got a shock: dark energy made a spectacular comeback, essential in explaining our observations.

Not only that, dark energy is not a minor cosmological player, but in fact dominates the budget of the universe. Dark energy makes up 70% of the energy in the universe, a percentage that will grow as the matter in the universe thins out but dark energy stays constant. These initial startling results have been backed up with other cosmological observations, so we cannot simply banish dark energy as the result of over-zealous astronomers staring too hard at their data. It is now an essential component of the *standard* big bang theory.

But we have very little idea what dark energy actually is! As with dark matter, there is plenty of speculation, but while dark matter is somewhat hemmed in by particle physics, dark energy is a true theorists' playground, where even the wildest ideas can't be decisively ruled out.

The first question must be: is there a substance that we know about that could act like dark energy in Einstein's equations? There is! As we mentioned above, dark energy may be a property of the vacuum.

This might sound strange if you are thinking about the vacuum as empty space. But our best theory of how matter behaves, known as *quantum field theory*, proposes that fundamental particles are really wiggles in fields that fill space and time. Electrons are wiggles in the electron field, up-quarks are wiggles in the up-quark field, and so on. Even when no particles are present, that is, the field is not wiggling in a particle-like way, the field is still there. If this *vacuum state* of the field has energy locked in, the vacuum itself will act as the source of cosmic acceleration.

But there is a problem. While vacuum energy has the universe-accelerating properties we need, the *amount* of vacuum energy predicted by particle physics grossly mismatches observations. How big a mismatch? A factor of 10^{120} or so. That's a bit awkward.

The source of this scientific tragedy could lie in the incompleteness of our theoretical ideas, with some as-yet-not-understood mechanism hiding the true value of the vacuum energy, or with our universe being just one in a multiverse, having received its cosmological constant through some roll of the universe-creating dice. To others, we need to discard vacuum energy and hunt for another form of energy or another field. Some think that we aren't correctly calculating the effect of structure in the universe on its overall expansion. Again, with a lack of observational clues, theoretical speculation runs wild.

So, the dark sector of the universe is a bit strange. The best ideas we have are that dark matter exists and is some kind of particle, and dark energy behaves in a way that is indistinguishable (with current data!) from vacuum energy. But there is the real possibility that both of these ideas are way off the mark. A Nobel Prize awaits.

Galaxy Formation and Galactic Satellites

Dark matter and dark energy are major puzzles, but for many cosmologists, the battleground on which they most often clash concerns galaxies. Let's explain.

Of all the ways that stars and gas could be arranged in the universe, why galaxies? To answer this question, modern cosmology increasingly looks to synthetic universes that are built inside a computer. The numbers inside the computer represent small (on a cosmic scale!) parcels of matter in an expanding space that feel the mutual gravitational pull of all the other parcels. In these numerical worlds, we can watch as small lumps of matter attract other lumps and grow, forming into galaxies, and galaxies into clusters and groups. Between the galaxies, matter forms thin filaments that are still deciding which galaxy they want to join, leaving extensive empty regions known as voids (recall Figure 6.6).

Beginning in the 1970s, astronomers have been compiling extensive surveys of galaxies, mapping out the three-dimensional distribution of matter in the cosmos. Figure 10.1 compares slices of the real universe with slices of a simulated universe. The statistical resemblance is impressive, providing good evidence that the standard big bang theory is a reasonable description of the universe.

As we watch the birth and growth of a large galaxy like the Milky Way, we see that it gains mass by collecting intergalactic gas and eating smaller galaxies. There is a hierarchy of galaxies, and the big ones get to the top by eating their underlings. This sounds more gruesome than it is: stars gather unscathed into a disk or bulge, and gas collides and pressurizes, often producing a new generation of stars.

So, how many of these small galaxies are out there? And what are their properties? This is where the fun begins.

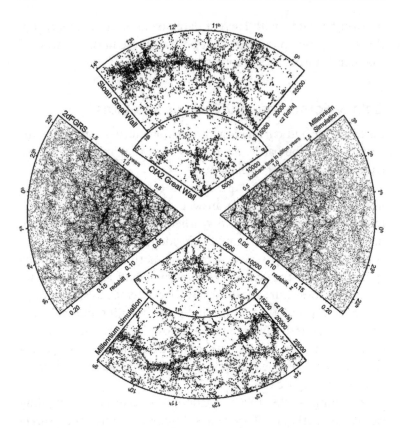

Figure 10.1. Observations of galaxies (left and top), and corresponding
sections of a universe in a computer known as the Millennium Simulation (right
and bottom). The left slice shows the 2-degree-Field Galaxy Redshift Survey
(2dFGRS), and the right shows a comparable simulation slice, taking into
account the fact that closer galaxies are easier to see. The top shows a piece of
the Sloan Digital Sky Survey, and the bottom is its comparable simulation slice.
There is obvious randomness in the patterns, but the overall statistical
clustering matches quite well. (Image credit: 2dFGRS, SDSS, Millennium
Simulation/MPA Garching, Volker Springel, Gerard Lemson and the Virgo
Consortium.)

By the end of the 1990s, our synthetic universes were detailed
enough to resolve the cosmic environment of big galaxies like
the Milky Way. The simulations showed that our galaxy should
be surrounded by hundreds or perhaps even thousands of dwarf
companions, buzzing like satellites around the Earth.

On the sky, these small *dwarf* galaxies are faint and difficult to detect. But as we looked harder with bigger telescopes, we had a problem: the dwarf galaxies weren't showing up. We found a handful of satellites at first, and now about 60 are known in the local group of galaxies. But the hundreds of predicted galaxies simply are not there, creating the *missing satellite problem*.

Is this the death-knell for the standard big bang theory? Not quite, as it has one card left to play: complex gas physics.

Earlier generations of simulations only modelled the effect of gravity, which effectively treats all the matter in the universe like dark matter. This significantly simplified the problem for the computers available at the time. So, these simulations didn't actually predict hundreds of *galaxies* – dark matter, stars, and gas – around the Milky Way. They predicted hundreds of collapsed balls of dark matter (known as *haloes*); it was assumed that they were big enough to host gas and stars, making a dwarf galaxy.

This assumption can be tested by including the complex gas physics in a new generation of simulations, made possible by improvements in computing power. These *hydrodynamical* simulations add more physics: gas will heat up, pressurize, cool down, form stars, and collapse into black holes. These are less important than gravity, but they are precisely the processes that create the light we can see.

How does this affect the missing satellite problem? The formation of stars in small dark matter haloes is significantly less efficient than in larger galaxies, meaning that there are too few stars for us to see them. Firstly, massive stars burn rapidly through their nuclear fuel before exploding in violent supernovae. In a large galaxy, the blast is absorbed, and the ejected material is recycled into the next generation of stars. But in a small halo, the explosion slams into surrounding gas and begins to drive it out. If the shock of multiple supernovae is sufficient to drive out much of the interstellar gas, then star formation will be completely shut down. Secondly, small galaxies are less able to shield their gas from intense intergalactic light. If the gas in a small galaxy is too hot, bathed in radiation, then it is too pressurized to collapse into stars. Once again, small galaxies fail to make stars.

So, within the standard big bang theory, there are hundreds of dark matter haloes out there, orbiting our and every other large galaxy, but, being deficient in gas and stars, they are invisible.

This explanation hasn't convinced all cosmologists. Perhaps dark matter doesn't sit still after it is born, but buzzes at a certain temperature. This *warm dark matter* moves fast enough that it can escape the weak gravitational pull of a small halo. In such a picture, there aren't hundreds of small dark matter haloes; these were erased as their particles dispersed.

So, we have two potential solutions. But is either one correct? We need a different way to test our ideas. In the past decade, the population of dwarfs has raised another puzzle, known as the *too big to fail* problem. It goes like this: if we see any of the small dark matter haloes, we should see the biggest ones. They are best able to hold onto their gas and make some stars for us to see. So, our simulations should at least be able to predict the number of large satellites. The dwarf galaxy population of the Milky Way is dominated by one large satellite, the Large Magellanic Cloud (LMC). But the simulations lead us to expect that we should see several satellites at least as large as the LMC. But we don't.

Again, debate over this large dwarf (so to speak) deficiency is ongoing, but the finger is again pointed at the usual culprit: complex gas physics. Galaxies of different masses aren't perfect scale models of each other, and their formation and evolution can be messy. The largest galaxies (by mass) aren't always the brightest at any given time.

An extraordinary amount of effort and computer time is being poured into refining computer simulations of galaxy formation to include more and more realistic physics. The hope is that, with enough computing power churning through the mathematics, our synthetic universe, dominated by dark energy and dark matter, will be a perfect match to the real one.

But is "gas physics" just a get-out-of-jail-free card, allowing us to hide significant problems behind complicated physics? Observers aren't just waiting around for the theorists to make the perfect simulation. They're looking for yet more ways to test our ideas.

Over the past decade, another problem with the population of satellite galaxies has come to light. We might expect these hanger-on galaxies to be scattered about their more massive hosts, if not perfectly randomly, then maybe resembling a slightly squished or slightly elongated ball. This what we see in our synthetic universes.

For most galaxies in the universe, the dwarf galaxy population is too faint to detect, so we can't check this prediction. Only the nearest galaxies are close enough for our telescopes to spy their dwarf populations in detail, including our own Milky Way and the Andromeda Galaxy, a mere 2.5 million light years away.

Hints that things were not quite right with our Milky Way's dwarf satellites were first noticed in the 1970s. In an influential study, Cambridge astronomer Donald Lynden-Bell wondered why the majority of the Milky Way's companions appeared to sit on a great circle over the sky.[1] Such a configuration suggests that the satellites are arranged in a plane, orbiting coherently.

Deeper surveys have uncovered more members of the dwarf population, including the spectacular Sagittarius Dwarf Galaxy, caught in the act of being devoured by the Milky Way. It's a messy meal: stars that have been stripped from the Sagittarius Dwarf wrap our entire sky, and in a few billion years, they too will be lost in the forest of stars in the outer halo of the Milky Way.

As more dwarf companions of the Milky Way were uncovered, the evidence that their orbits are far from random continued to grow. By the mid-2000s, we could draw on data from a variety of deep surveys of the halo of the Milky Way, using different telescopes in different hemispheres with different methodologies. The results were startling, revealing that the Milky Way's habitat is home to a varied population of globular clusters, dwarf galaxies, and stellar streams. Modern telescopes continue to probe our neighbourhood.

As more halo members have been identified, questions about their arrangement have remained. In 2005, Pavel Kroupa and his collaborators declared that Lynden-Bell's result from two decades earlier had been strengthened by new data.[2] Labelled the Vast Polar Structure (VPoS), Kroupa claimed that this plane of dwarf galaxies directly challenges the standard big bang theory, as similarly coherent structures were rarely spied in synthetic universes.

Kroupa argued that dark matter is the culprit, and that we should instead suppose that gravity is different from what Einstein taught us, at least on very large scales. With this modification, you no longer need dark matter to provide an extra pull to hold stars in their orbit. How does this arrange for dwarf galaxies to form in a flat(ish) plane? In Kroupa's universe, dwarf galaxies are produced by interacting large galaxies, which throw out streamers of gas that condense into dense knots of stars. We actually observe these *tidal dwarfs* in colliding galaxies, but in the standard big bang theory they are accompanied by no dark matter and so rapidly disperse. For Kroupa, the extra pull of gravity holds them together. The VPoS was formed when the

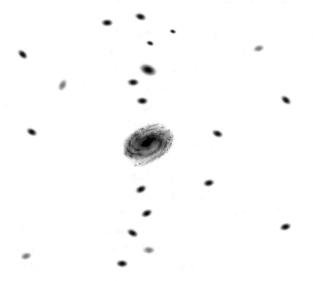

Figure 10.2. A large galaxy, like our own Milky Way, is surrounded by a swarm of dwarf galaxies. These are not randomly scattered, but some, quite unexpectedly, sit on a narrow plane that appears to be coherently rotating.

Milky Way encountered another large galaxy, the paths of the galaxies prefiguring the preferred plane of the satellites.

Kroupa's view is in the minority. Part of the reason is that the apparent plane of dwarf galaxies in the halo of the Milky Way is not widely seen as a challenge to the standard big bang theory: perhaps it's simply a statistical fluke, or some as-yet poorly understood effect of gas physics and star formation, or maybe better data from satellites such as the ESO Gaia mission will see the plane dissolve into a more uniform distribution. This may well be the case, but, as the evidence currently stands, the unexpected discovery of the coordinated dance of dwarf galaxies in the halo of the Milky Way is, at least, intriguing.

The interesting thing is that the Milky Way is not alone in possessing a coherent plane of orbiting dwarfs.

Over the past few decades our ability to map huge swathes of sky has significantly improved: more sensitive electronic detectors, larger telescopes, more advanced methods of overcoming noise and analysing data. As the faint sky has come into view, we have been able to hunt out dwarf companions around the Andromeda galaxy; more than 30 are now known to inhabit Andromeda's neighbourhood.

In 2014, the painstaking work of measuring the brightnesses, distances, shapes, and velocities of these galaxies had reached sufficient precision that we could answer the question of how the dwarfs of Andromeda are arranged. Geraint was directly involved in this effort. The results were spectacular: half of the dwarf population appear to lie on a remarkably narrow plane (for an astronomer!), 36,000 light years thick, but 1,200,000 light years in extent. Such a thin plane is quite unexpected, but that is only half the story.

When astronomers considered the motions of the dwarfs, the results were equally startling: those north of Andromeda heading towards us (relative to the motion of the Andromeda Galaxy itself), while those to the south were moving away. This suggests this entire plane of galaxies are orbiting in a coherent, organized fashion!

One plane of galaxies might be a coincidence, but two starts to look like a conspiracy. This was not predicted by any galaxy formation simulation before it was discovered. Is it possible that the physics of the standard big bang theory is enough to explain the data?

It seems unlikely, on face value. Remember that the dominant mass within a dwarf galaxy is not the stars we can see, but the dark matter we cannot. As a result, the stars are mostly irrelevant to the pull of gravity on the dwarf galaxy: to make the thin planes, we need to guide their dark matter.

What could do the guiding? We've already mentioned that galaxies aren't scattered at random; they assemble at the intersection of vast filaments of matter into groups and clusters. Could these filaments, which are seen in every galaxy formation simulation, guide dark matter more strongly than we have thus far suspected? In current simulations, this seems unlikely because the filaments are thicker than galaxies! Placing dwarf galaxies on fine planar orbits is akin to a gorilla repairing a watch; not impossible, but difficult to imagine.

At this stage, we simply do not have the answer. Be assured: these problems are not being ignored, let alone deliberately swept under the rug. Papers, with new calculations and new data, are being published every day. This is a cosmology battleground; what does your cosmological model have to say about dwarf galaxies and their arrangement?

Redshift Drift

The testing of the big bang theory continues. An important factor that drives more stringent observational tests is the continual improvement of technology: we can see deeper into the night sky, over wider areas, and measure what we see more accurately. This is not due simply to an increase in the size of telescopes, though this is very important, but also thanks to new instruments for analysing the light that is captured. These advances are allowing data to catch up with some long-known predictions of the big bang theory.

We have seen throughout this book that the redshift of distant galaxies is a key fact about our universe. The big bang explains redshift via the expansion of space, governed by Einstein's equations of gravity. Remember, redshift measures the fractional change between the wavelength of light emitted by a galaxy and the wavelength of the light as it arrives on Earth. If, while the light is on its way, the universe expands to double the size, we expect the light to arrive with twice the wavelength.

The rate of expansion of the universe changes, thanks to the pull and push of its contents. The expansion rate today is different from that yesterday, and it will be different again tomorrow. This means that a photon that arrived on Earth yesterday will have been stretched by a slightly different amount than one received today. If we keep watching a distance source of photons, whether a galaxy, quasar, or supernova, we would expect the observed wavelength of received light to *drift* over time, and at a rate that is predicted by the big bang theory.

The first person to investigate the redshift drift was Allan Sandage, a giant of twentieth century astronomy who dedicated his career to measuring the expansion rate of the universe. In 1962, he was able to calculate the magnitude of the redshift drift: around one part in 10^{13} per day! This tiny amount was well beyond the abilities of instrumentation at the time.

In light of its practical impossibility, Sandage's idea was lost to the mists of time, until it was rediscovered by Avi Loeb in the late 1990s. Loeb, apparently unfamiliar with the earlier paper of Sandage, made an additional realization: this idea's time was close. This *Sandage–Loeb test*, as it has now been dubbed, was within the reach of technological advances. As astronomers had strained to find planets outside our solar system, using tiny wobbles in absorption lines as the planet and star orbit one another, they had pushed their detectors to new heights of precision. Tiny shifts in wavelength, the result of the star wobbling in its orbit by as little as 1 km/h, can now be measured.

When measuring redshift drift, the longer we can measure, the easier it is to see the drift. However, even with foreseeable technology, observations over days and months will not be long enough. We'll need to make consistent, stable observations over decades, coming back to the same objects time and time again to build up the signal. We'll need the source of light to emit stably over those decades, but this shouldn't be too difficult for an entire galaxy and, in any case, we can cross-check by measuring several independent spectral lines. It will also help if the objects we observe are very distant, but that means they will also be faint. We'll need large telescopes, as well as highly advanced detectors.

These are coming, but maybe not until after we have retired from astronomy; not that academics ever really retire, as once they are released from the burdens of teaching, supervision, grant writing, and dealing with boundless university administration, they can actually get some research done. We may still be waiting for the first tentative detection, and then for further confirmation as more and more observations add to the signal and reduce the noise. This may be another feather in the cap of the big bang theory, or a thorn in its flesh. Time will tell.

Looking for Lithium

In Chapter 7, we discussed the elemental abundances of the cosmos. We saw that the big bang makes predictions for four of the lightest elements: helium-4, deuterium, helium-3, and lithium-7. For the first three, these predictions are remarkably accurate: helium-4 makes up 25% (by mass), deuterium makes up one nucleus out of 40,000, and helium-3 is one nucleus out of 90,000.

However, the big bang also predicts the amount of primordial lithium-7 and gets it wrong. We observe that one nucleus in every 6 billion is lithium-7, but the big bang predicts a value that is three times larger. This problem – known, creatively, as *the lithium problem* – has been puzzling cosmologists for a few decades, but has become particularly acute as measurements have become more precise.

We have a theory that has made a wrong prediction, and so *something* has to give. But the theory also makes some impressively correct predictions, so we don't want to be too hasty. "These observers aren't always correct, you know?", the big bang can say, curtly, "so I'm not gonna crack." Will a small correction to our ideas be enough, or do they need a radical overhaul?

Three types of solution to the lithium problem have been considered by cosmologists. The first is that something is amiss in our calculation of the cosmic oven. Perhaps our understanding of the nuclear physics involved has an as-yet unnoticed flaw or incompleteness that is throwing off the big bang's predictions. Even a decade ago, this was a viable option. However, more precise measurements of the relevant reactions by nuclear physicists have shut down this escape route.

Secondly, perhaps measurements of the cosmic lithium-7 abundance are systematically underestimating the true value. Recall that, to make this measurement, we look for somewhere in the universe that is untouched by stellar burning and other nuclear processes. We try to observe the wavelengths of light that indicate the presence of lithium-7. Whereas for deuterium, for example, we could probe pristine clouds of gas in young galaxies, for lithium-7 we rely on seeing these atoms in absorption in the outer atmospheres of stars. We look for stars that have relatively small amounts of the products of stellar nucleosynthesis. The most pristine stars all seem to have lithium abundances of 1.6 parts in 10^{10}. Perhaps there is some process that is using up lithium in these stars, but even then, it is a bit strange that there are no stars with abundances near the primordial value. Astronomers are attempting to measure lithium-7 in other cosmic environments; a measurement of the interstellar gas of a nearby dwarf galaxy known as the Small Magellanic Cloud gave a value that is pleasingly close to the primordial value. This escape route remains viable.

Finally, there could be exotic particle physics processes that affect the production of lithium-7 in the early universe. For example, adding an extra type of particle that interacts strongly (but not too strongly) with protons and neutrons could change the amount of lithium-7 without affecting the successful predictions of helium and deuterium. That may seem too easy, but in fact it is quite difficult to find a theory that will make just the right changes to the early universe.

We await better data and better theories. For the moment, the problem is simply unsolved. Cosmologists haven't taken the first available escape route and declared that everything is fine with the big bang. We have a problem. If your theory of the universe can explain *all* the cosmic abundances, including lithium-7, then it will have scored a victory over the big bang.

Primordial Neutrinos

A large proportion of this book has discussed the cosmic microwave background, radiation left over from the big bang that is visible across the entire sky. Whilst this radiation was unimaginably hot during the universe's first few minutes, it has cooled today to a chilly 2.725 K, just above absolute zero. Before the onset of nucleosynthesis, this radiation interacted energetically with protons, neutrons, and electrons, blasting apart any nuclei that attempted to form. After the first few minutes, the background radiation was still energetic enough to keep electrons from binding to nuclei to form neutral atoms. Only after 380,000 years was this sea of photons cool enough for neutral atoms to form, leaving the radiation to stream freely.

But in the initial moments, the protons, neutrons, and electrons were not the only particles colliding with the radiation: there were also the *neutrinos*. The neutrino is the strange cousin of the electron in the standard model of particle physics. Relative to other particles, they have only a tiny sliver of mass and no electric charge, so they do not feel the presence of the energetic photons. Through the nuclear weak force, they can interact with protons, neutrons, and electrons.

Just like photons, we can talk about the temperature of a population of neutrinos, which tells us the average energy of an individual neutrino. The matter in the early universe was in *thermodynamic equilibrium*, with a continuous flow of energy between the various populations of particles. While neutrinos are exchanging energy freely with other types of particles, they have the same temperature.

This continued until a critical point, which happened about 1 to 10 seconds after the birth of the universe. At this time, the expansion of the universe had diluted the cosmic soup so much that neutrinos lost touch with everything else. The amount of time a typical neutrino had to wait before running into a proton or an electron became longer than the age of the universe, and it only got worse. If a neutrino needs to wait one hour before happening upon an electron, but there has only been one minute since the beginning of the universe, then we've got a problem. And if we wait one hour, only to find that the neutrino now needs to wait one day, then the problem is getting worse. Neutrinos are left alone. Cosmologists call this *freeze-out*.

Soon after freeze-out, the annihilation of electrons with positrons pours some extra radiation into the universe. But the solitary neutrinos, because they won't interact with the photons or with anything else, miss out on their share. Energy gets shared around in thermal equilibrium, but they aren't in thermal equilibrium with the other particles in the universe any more. As a result, our universe's photon background is hotter than its neutrino background by a precise factor of $(11/4)^{1/3}$, which amounts to about 40%.

Both populations cool as the universe expands. Today, the expected temperature of the neutrino background is 1.95 K. If we could detect these neutrinos and measure their temperature, this would be a vital test of the big bang theory.

However, detecting neutrinos is not easy at the best of times. They only interact via the weak nuclear force, so arranging for them to trigger your detector is difficult. At this moment, neutrinos are flowing from the centre of the Sun to the Earth, with around a hundred billion passing through every square centimetre of your body every second. The chances of any one of these neutrinos interacting with one of your atoms, or even with the Earth's atoms, is tiny. Almost all will stream through as if the Earth was completely transparent.

To build a neutrino detector, you're going to need something big, and a lot of patience. For example, Japan's Super-Kamiokande uses 50,000 tonnes of ultra-pure water, while the IceCube experiment monitors a cubic kilometre of ice in Antarctica, watching intently for that exceedingly rare moment when a neutrino strikes a proton in a nucleus, firing out a high-speed electron that emits a burst of light. As our technology advances, neutrino astronomy is becoming a reality, allowing us to see directly into the heart of the Sun, whose nuclear reactions are a prodigious source of neutrinos.

What are the prospects for detecting the primordial neutrino background? These neutrinos, while born in the fires of the big bang, have cooled to exceedingly low energies today, many millions of times less energetic than the neutrinos streaking from the solar core. This lack of energy means that the chance of them interacting with an atomic nucleus is virtually zero. Given our current understanding of neutrinos, exceptional advances in technology will be needed before we can directly detect the neutrino background.

But this is not the end of the story. We can't directly detect them, but these neutrinos still have an impact on the universe. Every form of energy in the universe affects the expansion of space, which in turn affects the cooling of the cosmic ovens during nucleosynthesis. Adding or subtracting some neutrinos from the usual big bang story would affect the abundances of the elements, which we discussed in Chapter 7. When cosmologists calculate the expected abundances, they usually start with the standard model of particle physics, which says that there are three types of neutrinos, and go from there. But we can turn the problem around and instead pose the following question: starting with the cosmic chemical abundances, how many types of neutrinos are there?

Now, because our data about the abundances isn't perfectly precise, our inference about the number of types of neutrinos isn't either. The answer we get is: between 1.9 and 4.5, with the most likely value being about 3.1; in the equations, we can treat the number of neutrinos as continuous, so a fraction of a type of a neutrino does make sense. This is a nice result: the neutrinos we *need* to get the right rate of cooling in the early universe is, plus or minus some loose change, equal to the neutrinos we *get* from particle physics. If the cosmic oven had required zero neutrinos, or 100 types of neutrinos, then we would have a serious problem on our hands. As it stands, we have a nice piece of indirect evidence for cosmic neutrinos.

Neutrinos play another role in our cosmos, affecting how structure forms. When gravity is pulling clumps of matter together to make yet larger and larger clumps – and so on until galaxies – it matters how the matter is moving. A cloud of atoms, just floating calmly in space, will be pulled by its own gravity and begin to collapse on itself. But neutrinos don't sit still! Like photons, they are always moving, and very close to the speed of light. As a result, neutrinos tend to hold back the progress of structure formation, smoothing out lumps and bumps by freely streaming between them.

By the time the cosmic microwave background is released from matter and streams towards us, its pattern of slightly hotter and colder patches contains the imprint of this neutrino smoothing. It is also imprinted on the pattern of galaxies we see across the universe. As above, we can start with our observations of the universe and ask: how many types of neutrinos are there?

Again, observations give us an answer with a finite degree of precision: between 2.66 and 3.33, with the most likely value being about 2.99. That's nice. It's indirect evidence, but it's quite satisfying.

If we could just see those neutrinos directly, that would be even nicer. Anyone? We're all ears.

Unmaking Antimatter

As we discussed in Chapter 7, the early universe is a melee of energetic particle processes. If a certain process – such as pulverizing a nucleus into its constituents – needs a certain amount of energy, then there is some point in the early universe when each particle in the universe has enough.

There is one particular process that has received a lot of attention from cosmologists. Remember $E = mc^2$; you can make mass, if you have enough energy. At some point in the early universe, particles of light can slam into each other and create matter. Specifically, they create *matter–antimatter* pairs. Each particle in the universe has an antiparticle: electrons have positrons, protons have antiprotons, neutrinos have antineutrinos, and so on. This is not some zany, speculative theory; antiparticles have been directly observed.

Before the universe was one second old, the universe was hot enough that photon collisions could create an electron–positron pair. Before the universe was 10^{-5} seconds old, photon collisions could create a proton–antiproton pair. So, the very early universe was awash with all kinds of fundamental particles, transforming into each other.

When we run the clock forward, we notice something odd. When we measure the properties of matter and antimatter, we find that they are either equal, or equal-and-opposite. For example, electrons and positrons have equal masses, and equal-and-opposite charges. They are mirror images of each other. With matter and antimatter being created so freely, and no relevant difference between their properties, we would expect equal amounts of each to exist in those first moments.

But now keep your finger on the play button, and watch the universe expand. Soon, the universe is too cool to make proton–antiproton pairs. A fraction of a second later, the universe is too cool to make electron–positron pairs. These particles find each other and *annihilate*, which is what particle physicists call the reaction "particle hits antiparticle and makes two photons". The reverse reaction, the creation of a particle and antiparticle, doesn't happen because there isn't enough energy in the photons any more.

Is there any antimatter in the universe today? Very little, as far as we can tell. There is very little on Earth, very little in the solar wind that sweeps past Earth, very little on the Moon, and no sign of widespread annihilation as matter regions meet antimatter regions anywhere in the universe. So, it all must have annihilated in the early universe.

But if the very early universe contained equal numbers of protons and antiprotons, and equal numbers of electrons and positrons, then when they all pair off and annihilate, won't there be none left? Why does the universe have any matter at all, rather than just photons?

In our universe, there are approximately a billion photons for each particle of matter. That's a small fraction of matter, but still a whole lot more than none. It means that, before the wholesale annihilation of matter–antimatter in the first moments, there were roughly a *billion and one* protons for every billion antiprotons. When a billion protons and a billion antiprotons paired off and created light, there was one proton left over, unpaired. The story is the same for electrons as they pair off with positrons. You, this book, the Earth, the Milky Way, and everything we can see is made of those billion-and-first protons and electrons.

It is sobering to think that our existence owes itself to an imperfection in the early universe of one part in a billion! Without it, the universe would be little more than a featureless sea of ever-cooling photons and neutrinos. How did a symmetric universe manage this?

In 1967, the Russian physicist Andrei Sakharov identified the necessary conditions for creating an overabundance of matter over antimatter. In short, we need matter and antimatter to be imperfect mirror images in some respects, and a period in the universe that exploits these differences. Now, the interesting thing is that these asymmetries aren't purely theoretical. We have observed processes in particle accelerators that treat particles and antiparticles differently. The standard model of particle physics has the features that Sakharov identified.

But alas, they aren't strong enough to create the matter–antimatter asymmetry we observe in our universe. And so cosmologists and particle physicists are busy proposing new ideas for processes that will create the observed asymmetry. This is far from a trivial problem, but viable ideas can be found in academic journals. Sadly, none of the workable models rely entirely on known physics. To test these models, we would need to see the processes they propose at work in particle accelerators, or in the universe we can observe today.

Or, perhaps you can do better. If your idea about the universe can explain the predominance of matter over antimatter in the universe without having to propose new, untested particle physics processes, then it will score a major win over the big bang. Just saying that the universe started off with one-in-a-billion matter particles to photons isn't good enough. That's just restating the data. Any theory could do that and be as good as your theory. You need to do better than the raw data, uncovering the secret pattern that tells us why the universe is what it is. The door is wide open.

Primordial Gravitational Waves

As fond as we are of neutrinos and positrons, seeing the big bang itself would really be something. Or, at least, seeing into those mysterious, extreme conditions in the first moments: immense densities, unimaginable temperatures, as-yet unknown laws of nature, and who knows what else! Inflation ... maybe.

The physics of inflation is poorly understood, but that hasn't stopped hundreds of papers being published on the topic, proposing models and ways they might be tested. One thing that many models of inflation have in common is that, as this period of super-expansion comes to an end, the oscillations of the energy of inflation are imprinted on the spacetime of the universe as a sea of gravitational waves, streaming through the cosmos like ripples on a pond. Just as left-over radiation is seen as the cosmic microwave background, our spacetime ripples with a background of gravitational waves.

Whether or not inflation happened, these primordial gravitational waves are very exciting for cosmologists. Recall from Chapter 5 that the cosmic microwave background (CMB) represents the furthest that we can see *with light*. A photon in the universe, before 380,000 years after the big bang, is in a fog. The light will not travel directly to us, and so it cannot *directly* carry information about the conditions in the universe before the CMB was released. Think about a very cloudy day – you can see around you because of the sunlight that comes through the clouds, but the light is so scattered that you can't point in the direction of the Sun.

By contrast, gravitational waves will happily pass through the fog of protons and electrons in the early universe. This is like hearing a voice in the fog: you might not be able to see the speaker, but you can understand them. If gravitational waves were generated in the very early universe, they will travel unimpeded to us – stretched by the expansion, but not distorted or scattered or blocked. If we can detect them, we will hear the first moments of the history of our universe.

But that's where the problems begin. In 2016, the Laser Interferometer Gravitational-Wave Observatory (LIGO) announced the first ever direct detection of gravitational waves. Just a year later, LIGO's leaders Rainer Weiss, Barry Barish, and Kip Thorne won the Nobel Prize. LIGO's technology is astounding, bouncing laser light through kilometres of vacuum tube to measure immensely small distortions in spacetime.

LIGO, and its sibling detectors that are being planned and built, can only hear gravitational waves of a particular range of wavelengths. This is analogous to only hearing sounds with a certain range of pitches – the human ear can hear about ten octaves, while LIGO can hear about seven. Just as importantly, they are only sensitive enough to hear intense bursts of waves from some of the most violent events in the universe: merging black holes and neutron stars.

Because of the extreme stretching of waves from the very early universe, any gravitational wave background is probably too faint and too low in pitch (too long wavelength) to be detected by current technology. Future detectors, including a planned network of detectors in space called LISA (Laser Interferometer Space Antenna) will get closer to this limit, but if they hear anything, it is more likely to be the background hum of orbiting pairs of black holes and neutron stars, drowning out any primordial signal.

However, these are guesses, based on the best information we have but far from certain. As always in science, we don't know for sure until we look. And if that means bouncing lasers between space probes that are 5 million kilometres apart, then we'll try that too, hopefully not too far in the future.

But there is another way! In Chapter 8 we discussed the imprint of gravitational waves on the CMB. Before 380,000 years after the big bang, the universe was a hot plasma of protons and electrons being squeezed and stretched by the primordial gravitational waves coursing through spacetime. We have a chance to glimpse this signal by looking at the polarization of the CMB.

As we saw previously, cosmologists had a false start with measuring this signal, thanks to the (now discredited) BICEP2 announcement of the measurement of primordial gravitational waves in 2014. The BICEP2 team had fooled themselves into thinking they detected a signal from the very early universe, whereas they had actually measured dust in our own Milky Way.

For the moment, primordial gravitational waves await detection. Their discovery wouldn't so much test the big bang theory as give some guidance as to what may have come in those first moments.

The Birth of the Universe

Now we come to the big one: the birth of the universe. We've spoken a lot about our understanding of our cosmic evolution, about how matter and light flowed and changed as the universe expanded and cooled. We've also discussed how, as we peer deeper and deeper into the past, we eventually reach a wall of our own ignorance. Suppose you want to know what was happening in the universe when it was one second old. The big bang theory, informed by the CMB, predicts that the temperature of the universe was 10^{10} K. So, you'd better know what matter does when it's that hot.

And, we do. Particle physics experiments have thoroughly explored matter at these temperatures. But there is a limit to our experiments: when the universe was about 10^{-14} seconds old, almost every particle in the universe had more energy than any particle in our experiments. Our theories can predict what happened before, including speculation about inflation and the unification of the forces of nature, but we are beyond what we can check with data. Our understanding of the laws of physics becomes more and more uncertain.

At yet earlier times, we hit an even harder wall. As we have discussed, gravity just doesn't behave like the other forces. Einstein described the force of gravity as a consequence of the bending and stretching of space and time; by contrast, the other fundamental forces (electromagnetism, the strong nuclear force, and the weak nuclear force) are described in terms of interacting quantum fields that live in space and time. So, one set of mathematical equations describes the action of gravity, whilst a completely different set describes the quantum forces. In their own comfort zones, each theory works marvellously. But we don't know how to describe all four in all circumstances. We don't have a theory that tells us what happens when the fundamental forces are all acting together, and, in particular, when gravity is showing its quantum side.

We want a *Theory of Everything*. This is a slightly over-enthusiastic way of saying that we are looking for a single mathematical description of all of the forces of nature. The search for this ultimate theory has been underway for almost a century. Einstein was trying to unify electromagnetism and gravity on his deathbed, and today a battalion of physicists is trying to advance today's more promising ideas, such as M-theory or quantum loop gravity. At the moment, we don't know if these are the path to enlightenment, or simply more complicated blind alleys.

What does this have to do with the birth of the universe? Well, until we can understand all of the fundamental forces firing at full capacity, we can't leap over that theoretical wall that is stopping us from peering into the past.

We at least think we know *when* this wall occurs. It is some-where around the Planck time, about 10^{-43} seconds after the birth of the universe. But isn't this a contradiction? How do we know that this point is 10^{-43} seconds after the beginning if we don't know what happened before this time?

The answer is that we are being a bit sloppy. We can ignore the quantum forces, and just use Einstein's theory of gravity. This allows us to overcome the theoretical wall to see *something* before the Planck time. In this universe, as we look back in time, the density and temperature continue to grow higher and higher, until we hit a point, the birth of the universe, where the temperature and density soar to infinity.

In a no-quantum-Einstein-only universe, this starting point is the start of everything, of all space and all time. It's a day without a yesterday, the beginning of physical time itself.

But, obviously, we've left out something important! What should we expect if and when we can break through the barrier of the Planck wall, theory of quantum forces and gravity in hand? Perhaps the beginning remains. It is worth remembering that no cosmologist formulated a theory of the universe with the express intention of *adding* a beginning. Cosmologists proposed models that described the universe, and later *discovered* a beginning in their model. By contrast, many cosmologists have tried to remove the beginning, and failed.

But perhaps this new theory will reveal that our universe is just one in a long line in a cycle of cosmic births and deaths, or that we are one of many daughter universes from a mother universe, itself one of myriad daughters. Or we might find higher dimensional objects of which we are merely a part, with universes floating and colliding in an uber-universe. Or, as seems likely, something we haven't even thought of yet.

The birth of our universe, or even whether our universe had a birth, remains a mystery to cosmology. Many of our clues point to a beginning, but not all, and none decisively. We find a theoretician's playground, with cosmologists often making grand claims about how the universe may have come to be. It's an exciting but crowded field, where speculation abounds, data is rare, but the potential rewards could be immense.

The big bang theory has a missing piece at the start. But there is a deep reason why the beginning of the universe is, and may always remain, a scientific mystery. Every theory of physics thus far proposed, from Newton's laws to fluid flows to quantum mechanics to Einstein's theory of gravity, tells us how the universe *changes*. When we apply these laws to *any* physical system, whether the Solar System or a hydrogen atom, we need to tell the equations how the system started off. In the lingo, we need to supply *initial conditions*. The theory doesn't tell you what they are. It can hardly be surprising, then, that no theory tells us about the start of the entire universe. The physicist John Wheeler makes this point forcefully:[3]

> *Never has physics come up with a way to tell with what initial conditions the universe was started off. On nothing is physics clearer than what is not physics: equation of motion, yes; initial position and velocity of the object which follows that equation of motion, no.*

It seems that, to understand the beginning of the universe, we not only need a new physical theory – we need a new *kind* of physical theory, one that says something about its own initial conditions.

So, if your idea wants to step into the limelight, and unveil the beginning of the universe, it will have a lot of explaining to do. Your theory needs to be mathematically formulated, so that it can make predictions about data. If it can do that, and also shed light on the beginning (or not) of our universe, then it will inform one of the biggest questions of existence.

Where to from Here

Throughout this book, we have considered the observational evidence that supports the standard model of modern cosmology, revealing a universe born in a fiery big bang (maybe), dominated by dark matter and dark energy (probably), and steadily cooling and forming stars and galaxies over the last 13.8 billion years (very likely).

If you are the proud owner and operator of your own cosmological model, one that is more logical, more sensible, and more coherent than the big bang, and your model has truly survived to this point, then well done! You have a contender.

We stress again that words and pictures are not enough. High-school mathematics is unlikely to be enough. Just pointing out problems and missing pieces of the big bang model is not enough. It has to be a truly scientific model, grounded in the language of mathematics, and open to all to scrutinize. Anyone with the requisite mathematical knowledge should be able to reproduce your calculations. This is a minimal requirement, without which you don't really have a theory.

In this last chapter we have discussed where the big bang theory struggles, where there are questions to be answered and phenomena to explain. This gives you the opportunity to strike. Can your model naturally address one or more of these questions?

Always remember, at its heart, science invites one-upmanship. Experimenters and observers want to prove the theorists wrong (especially a physicist by the name of A. Einstein), and theorists want to be one step ahead, predicting observations before they are made. If your theory can handle the truth – *all* the truth and nothing but the truth – then you are onto a winner. Look again, deep into your equations, and see if you can think of an experiment that will test your theory.

Finally, here are our 10 Steps to Beating the Big Bang and Building a Better Universe.

1. Know your enemy: Learn the mathematical model that undergirds the big bang theory.
2. Prepare your rival: Your theory of the universe needs to be precisely formulated, ready for anyone to use to predict data.
3. Olbers's paradox: Explain why the sky is dark at night.
4. Redshift and distance: Explain why the light we receive from galaxies is almost always redshifted, and why redshift increases proportionally to distance.

5. Fading light: Explain why supernovae fade more slowly, and why galaxies are dimmer, when their light is more redshifted.

6. Cosmic microwave background: Explain why the CMB has a near perfect blackbody spectrum at 2.725 K, and why the temperature of that spectrum increases with redshift.

7. Forest: Explain where the Lyman alpha forest comes from, why it is thicker at higher redshift, and why it is correlated with the positions of galaxies.

8. Cosmic oven: Explain why the universe (especially the parts untouched by stars) is 75% hydrogen, about 24% helium, and about one nucleus in 40,000 deuterium.

9. Do better: Explain or eliminate dark energy and dark matter, decipher satellite galaxies, solve the lithium problem, analyse the absence of antimatter, and unveil the beginning (or not) of the universe.

10. Publish: Present your theory – principles, equations, and predictions – concisely and clearly to the light of expert scrutiny.

Engage scientists through the right channels, be patient, and strike a blow for the revolution!

(Just be sure to give us at least *some* of the credit.)

Good luck!

ACKNOWLEDGEMENTS

This book arose through many conversations, both professional and not, on the nature of the universe and how we understand it. Much of it comes from interactions with the general public, at astronomy talks and events, where we are continually surprised by the people who turn out to listen to us drone on about the universe. We thank them all, young and old, those that listen, those that question, and those that challenge. We especially thank the children and teenagers, including those that ask the most piercing questions, because you really make us think. To them we say – don't simply "believe" what somebody says because they are older than you. Never be afraid to ask "why?"

Some colleagues questioned why we were writing this book, often followed by "where did you get the time?" But their conversations were invaluable, and we particularly would like to thank: Joss Bland-Hawthorn, Rodrigo Ibata, Bob Carswell, Ryan Cooke, Allen Hainline, Michel Bourdeau, and Mike Keas. We'd like to thank Gerard Lemson, Volker Springel, Daniel Eisenstein, and David Malin for providing some of the figures.

We thank Vince Higgs, the senior commissioning editor for physics and astronomy, and the staff at Cambridge University Press for their professionalism and support, and for taking another chance with our second book.

Luke thanks Geraint for having the idea for this book in the first place and inviting me to get involved. I am delighted to announce that a fraction of all sales of this book will go towards a speedboat for Geraint – we've nearly paid the deposit on a steering wheel. Luke is supported by a grant from the John Templeton Foundation. The opinions expressed in this publication are those of the author and do not necessarily reflect the views of the John Templeton Foundation.

Luke thanks his family for all their support. My little ones are six and nine now, but still want a high-five from Dad as they leave for school. Thanks for coming to some of my talks at the observatory, and for listening patiently even when you know that as soon as I finish, the 3D movie will start.

Geraint thanks Luke for embarking on this book-writing journey again. The last couple of years have been immense fun, and I feel like we have learnt an awful lot. A few more books, and we might get the hang of this publishing lark.

Family and friends, both near and far, really matter in life, so special thanks to Jonny 9-Fingers and Mat for the punctuated get-togethers that make life such fun, my parents Meurig and Yvonne and brother Ceri, and the extended family spread around the world, for always being warm and welcoming. To my boys Bryn and Dylan, you are my world, and a simple thank you cannot convey the love and happiness you bring into my life. And last, but definitely not least, my darling Zdenka, I can only quote the great Freddie Mercury – "you're my best friend".

ENDNOTES

Chapter 1

1 The idea of stretching space must be approached with some caution. While mathematically well defined, pushing the picture too hard can result in confusion. With some colleagues, we've discuss this point at length in: Francis, Matthew J., Luke A. Barnes, J. Berian James, and Geraint F. Lewis, 2007, "Expanding Space: the Root of All Evil?", *Publications of the Astronomical Society of Australia*, volume 24, issue 2 (arxiv.org/abs/0707.0380).

2 There are mathematical methods that help scientists assess if one model is a better description of observations than another. We recommend: *Bayesian Logical Data Analysis for the Physical Sciences: A Comparative Approach*, by Phil Gregory (Cambridge University Press, 2010).

3 Keas, Mike N., 2018, "Systematizing the Theoretical Virtues", *Synthese*, volume 195, issue 6. Note that we have simplified Keas's presentation of the Theoretical Virtues. Read the paper for his more precise formulation.

4 Source: Chandrasekhar, S., 1987, *Truth and Beauty: Aesthetics and Motivations in Science*, University of Chicago Press (pages 22, 52).

5 We note that some physicists feel that this obsession with beauty in physical theories has been taken too far. For a recent view, check out *Lost In Math: How Beauty Leads Physics Astray* by theoretical physicist Sabine Hossenfelder (Basic Books, 2018).

6 This epigram appears at the beginning of many manuscripts of Ptolemy's *Almagest*. See Gingerich, Owen, 1980, "Was Ptolemy a Fraud?", *Quarterly Journal of the Royal Astronomical Society*, volume 21, page 253.

7 Weinberg, Steven, 1967, "A Model of Leptons", *Physical Review Letters*, volume 19, no. 21.

Chapter 2

1 Chapman, Allan, 2013, *Slaying the Dragons: Destroying the Myths in the History of Science and Faith*, Lion Books.

2 This remark comes from a letter from Newton to a rival physicist Robert Hooke on 5 February 1676. It is sometimes alleged that Newton, by referring to "giants", was making fun of Hooke's short stature. However, the remark was in response to a compliment by Hooke, and there is no hint in the context of anything but approval of Hooke's work. Newton writes:

"At the reading of your letter I was exceedingly pleased and satisfied with your generous freedom, and think you have done what becomes a true philosophical spirit. … you defer too much to my ability for searching into this subject. What [René] Descartes did was a good step. You have added much several ways, and especially in taking the colours of thin plates into philosophical consideration. If I have seen further it is by standing on the shoulders of giants."

The complete letter can be found at The Newton Project: www.newtonproject.ox.ac.uk/view/texts/normalized/OTHE00101

3 Grant, Edward, 2007, *A History of Natural Philosophy: From The Ancient World To The Nineteenth Century*, Cambridge University Press.

4 Quoted in Ronald L. Numbers and Kostas Kampourakis (eds.), 2015, *Newton's Apple and Other Myths About Science*, Harvard University Press (page 24).

5 If you've heard that Aristotle just repeated old wives' tales and urban legends, read *The Lagoon: How Aristotle Invented Science* by Armand Marie Leroi (Penguin, 2015). "[Aristotle's] intimacy with the natural world shines from his works." (page 375).

6 "Virtually everything of Aristotle's theory of motion is still valid. It is valid in the same sense in which Newton's theory is still valid: it is correct in its domain of validity, profoundly innovative, immensely influential and has introduced structures of thinking on which we are still building." Rovelli, Carlo, 2015, "Aristotle's Physics: a Physicist's Look", in *It From Bit or Bit From It?*, edited by A. Aguirre et al., Springer International (arxiv.org/abs/1312.4057).

7 As pointed out by historian and astronomer Owen Gingerich, Kepler's laws "were not specially singled out and ordered as a group

of three until 1774", and furthermore he didn't call them *laws*.
That terminology started with Descartes and didn't really catch on
until Newton. Gingerich, Owen, 2011, "Kepler and the Laws of
Nature", *Perspectives on Science & Christian Faith*, volume 63, no. 1,
page 17.

8 Newton, Isaac, 1687, *Philosophiae Naturalis Principia Mathematica*
(Mathematical Principles of Natural Philosophy). Available online at
books.google.com.au/books?id=Tm0FAAAAQAAJ.

9 Was Newton correct in his deduction about the stability of his
universe? The net force on a given star in an infinite Newtonian
universe would be "infinity minus infinity", which is
mathematically undefined. The usual way of calculating motions in
Newtonian gravity will not work in an infinite universe – the
gravitational potential is infinite. An interesting paper by Frank
Tipler suggests that the infinite Newtonian universe is still
unstable, and will tend to collapse in an amount of time that
depends only on its density, not its size. In this way, it is like a
contracting cosmology in Einstein's theory of gravity. This suggests
that Newton was actually wrong in deducing that his universe
would be stable against collapse. Tipler, Frank J., 1996, "Newtonian
Cosmology Revisited", *Monthly Notices of the Royal Astronomical Society*,
volume 282, issue 1.

10 For a complete history of Olbers's paradox, see *Darkness at Night* by
Edward Harrison (Harvard University Press, 1989).

11 In the lingo, it would need a *fractal dimension* smaller than two.

Chapter 3

1 Available as a wall poster at www.fieldtestedsystems.com/ptspectra.

2 This is something of a simplification. Hubble had Cepheid
observations for some local galaxies, and for the rest assumed a
fixed intrinsic brightness for the brightest star in the galaxy,
writing, "The apparent luminosities of the brightest stars in such
nebulae are thus criteria which, although rough and to be applied
with caution, furnish reasonable estimates of the distances of all
extra-galactic systems in which even a few stars can be detected."
Hubble, Edwin, 1929, "A Relation between Distance and Radial

Velocity among Extra-Galactic Nebulae", *Proceedings of the National Academy of Sciences*, volume 15, number 3.

Also, in 1927 Georges Lemaître had assumed a linear velocity–distance relationship and calculated its slope (what is now called the *Hubble constant*) using similar data. But he did not plot the data, and so didn't show the relationship that we attribute to Hubble. Lemaître's original paper has been republished: Lemaître, Georges, 2013, "A Homogeneous Universe of Constant Mass and Increasing Radius Accounting for the Radial Velocity of Extra-Galactic Nebulae", *General Relativity and Gravitation*, volume 45, issue 8.

3 Hubble, E., 1929, "A Relation between Distance and Radial Velocity among Extra-Galactic Nebulae", *Proceedings of the National Academy of Sciences of the USA*, volume 15, issue 3, page 168.

4 Freedman, Wendy L. et al., 2001, "Final Results from the Hubble Space Telescope Key Project to Measure the Hubble Constant", *The Astrophysical Journal*, volume 553, number 1. In the plot, the grey wedge shows Hubble's Law for $H_0 = 75 \pm 5$ km/s per Mpc.

5 The squares are from Suzuki et al. (2012), and the triangles are from Betoule et al. (2014).

Suzuki, N. et al., 2012, "The Hubble Space Telescope Cluster Supernova Survey. V. Improving the Dark-Energy Constraints above $z > 1$ and Building an Early-Type-Hosted Supernova Sample", *The Astrophysical Journal*, volume 746, issue 1, id. 85.

Betoule, M. et al., 2014, "Improved Cosmological Constraints from a Joint Analysis of the SDSS-II and SNLS Supernova Samples", *Astronomy & Astrophysics*, volume 568, id. A22.

6 Pound, R. V. and Rebka, G. A., 1959, "Gravitational Red-Shift in Nuclear Resonance", Physical Review Letters, volume 3, page 439.

7 Mössbauer, Rudolf L., 1958, "Nuclear Resonance Fluorescence of Gamma Radiation in Ir 191", *Zeitschrift fur Physik*, volume 151, issue 2, pages 124–143.

8 To make this equivalence work, we just need to be a bit careful how we define velocity; see Emory Bunn and David Hogg, 2009, "The Kinematic Origin of the Cosmological Redshift", *American Journal of Physics*, volume 77, issue 8 (arxiv.org/abs/0808.1081). The advantage of the *Doppler view* is that it reminds us that light, in and of itself, doesn't have a wavelength. Its wavelength depends on the motion

of the observer. The advantage of the *stretching space view* is that it connects directly with the relevant mathematical formula: if space expands to be twice as large between emission and absorption, the wavelength of the light will be observed to be twice as long as when it was emitted. Another advantage of the stretching space view is illustrated by the following thought experiment. If the universe is finite in size, then light can travel around the universe and back to where it started. You could see the back of your own head, redshifted. In the stretching space view, space expanded while the light travelled around the universe. But the Doppler view requires you to be moving away from the back of your own head! See Roukema, Boudewijn F., 2010, "There Was Movement that Was Stationary, for the Four-Velocity Had Passed Around", *Monthly Notices of the Royal Astronomical Society*, volume 404, issue 1.

9 Stay updated at eventhorizontelescope.org for more detailed images in the future. Thanks to Natasha Moore, whose questions were a reminder that, until four days before this book's due date, we'd written, "However, quasars are too small for us to see the important details of the black hole and its disk."

10 Galianni, P., Burbidge, E. M., Arp, H., Junkkarinen, V., Burbidge, G. and Zibetti, S., 2005, "The Discovery of a High-Redshift X-Ray-Emitting QSO Very Close to the Nucleus of NGC 7319", *The Astrophysical Journal*, volume 620, issue 1, pages 88–94.

11 www.thunderbolts.info/tpod/2004/arch/041001quasar-galaxy.htm [accessed: 10 March, 2019].

Chapter 4

1 Blondin, S. et al., 2008, "Time Dilations in Type Ia Supernova Spectra at High Redshift", *The Astrophysical Journal*, volume 682, issue 2, page 724.

2 Blondin, S. et al., 2008, ibid.

3 The situation becomes more complex when the bulb is so far away that it appears point-like. For such images, all the light essentially hits one photo-receptor on your eye, so moving further away doesn't decrease the number of receptors that receive the light.

4 See also the papers by Sandage and Lubin below, and references therein.

Pahre, Michael A., S. G. Djorgovski, and R. R. de Carvalho, 1996, "A Tolman Surface Brightness Test for Universal Expansion and the Evolution of Elliptical Galaxies in Distant Clusters", *The Astrophysical Journal*, volume 456.

Lubin, Lori M., and Allan Sandage, 2001, "The Tolman Surface Brightness Test for the Reality of the Expansion. IV. A Measurement of the Tolman Signal and the Luminosity Evolution of Early-Type Galaxies", *The Astronomical Journal*, volume 122.

Sandage, Allan, 2010, "The Tolman Surface Brightness Test for the Reality of the Expansion. V. Provenance of the Test and a New Representation of the Data for Three Remote Hubble Space Telescope Galaxy Clusters", *The Astronomical Journal*, volume 139.

5 Sandage, Allan, 2010, ibid.

Chapter 5

1 Wild, J. P., 1974, "A New Look at the Sun", in *Highlights of Astronomy*, Volume 3, edited by G. Contopoulos, Dordrecht: D. Reidel.

2 McKellar, A., 1941, "Molecular Lines from the Lowest States of Diatomic Molecules Composed of Atoms Probably Present in Interstellar Space", *Publications of the Dominion Astrophysical Observatory*, volume 7, issue 6, pages 251–272.

3 Page 496 of Herzberg, Gerhard, 1950, *Molecular Spectra and Molecular Structure*, Princeton, NJ: D. Van Nostrand Company.

4 McKellar died in 1960, at the young age of 50. Nobel Prizes are not awarded posthumously, so sadly he could not have been a recipient.

5 Lemaitre, G., 1931, "The Beginning of the World from the Point of View of Quantum Theory", *Nature*, volume 127, page 706.

6 Simon Singh, "Dead of Night - the movie that changed the universe", www.theguardian.com/film/2005/jan/05/1 [6 January 2005, accessed 10 March 2019].

7 The question of whether radio observations really showed that the universe is evolving sparked a long-running feud between astrophysicist Fred Hoyle in one corner, and radio astronomer Martin Ryle in the other. Both were at the University of Cambridge,

with Hoyle head of the Institute of Astronomy and Ryle at the Mullard Radio Astronomy group. The two departments were on either side of Madingley Road in Cambridge, and the feud made this small road feel like the Atlantic Ocean. Over time, however, observations convinced the majority of astronomers that the past universe was distinctly different from today. For more on this history, see *Fred Hoyle: A Life In Science* by Simon Mitton (Cambridge University Press, 2011).

8 Noterdaeme, P., P. Petitjean, R. Srianand, C. Ledoux, and S. López, 2011, "The Evolution of the Cosmic Microwave Background Temperature: Measurements of TCMB at High Redshift from Carbon Monoxide Excitation", *Astronomy & Astrophysics*, volume 526 (arxiv.org/abs/1012.3164).

Chapter 6

1 Hazard, C., M. B. Mackey, and A. J. Shimmins, 1963, "Investigation of the Radio Source 3C 273 by the Method of Lunar Occultations", *Nature*, volume 197, page 1037.

2 This is a composite spectrum from the FIRST Bright Quasar Survey. The redshift of 2.5 was chosen to put the brightest line, Lyman alpha, in the optical window. Brotherton, M. S. et al., 2000, "Composite Spectra from the FIRST Bright Quasar Survey", *The Astrophysical Journal*, volume 546, issue 2, page 775.

3 Because the universe doesn't like being nice and neat for us, there is an exception. The rare *Wolf–Rayet* (WR) stars are known for their strong, wide emission lines. They are very massive stars that only last 5 million years, and only 10% of this time is spent in their WR phase: the emission lines happen when the star begins losing enormous amounts of mass, blowing at high speeds away from the star and so thick that we're not seeing the star directly. We're seeing its energetic *stellar wind*. Of all the stars we've observed in the Milky Way, only a couple of hundred are WR stars. We can measure parallax distances to WR stars, which we can't to quasars, so we know that they are in our galaxy.

4 Adelberger, K. L., C. C. Steidel, A. E. Shapley, and M. Pettini, 2003, "Galaxies and Intergalactic Matter at Redshift $z \sim 3$: Overview",

The Astrophysical Journal, volume 584, issue 1 (arxiv.org/abs/
astro-ph/0210314). The introduction to this paper is a wonderful
piece of writing and should be read by every up-and-coming
astronomer.

Chapter 7

1 Page 8 of Feynman, Richard P., 1964, *The Feynman Lectures on Physics*,
   Addison–Wesley.

2 Principe, Lawrence M., 2015, "Myth 4" in *Newton's Apple and Other
   Myths About Science*, edited by Ronald Numbers and Kostas
   Kampourakis, Harvard University Press.

3 Comte, Auguste, 1835, *Cours de Philosophie Positive* (volume 2,
   Nineteenth Lesson). Translation by Google. Available from
   www.gutenberg.org/files/31882/31882-h/31882-h.htm.
   Intriguingly, this quotation is significantly different in the
   1853 English translation *The Positive Philosophy of Auguste Comte* (Book
   2, Chapter 1), which seems to apply these doubts to planets,
   not stars.

4 Atreyaa, S. K. et al., 2003, "Composition and Origin of the
   Atmosphere of Jupiter – an Update, and Implications for the
   Extrasolar Giant Planets", *Planetary and Space Science*, volume
   51, page 105.

5 Asplund, M. et al., 2009, "The Chemical Composition of the Sun",
   Annual Review of Astronomy & Astrophysics, volume 47, issue 1,
   page 481.

6 Nieva, M. F. and N. Przybilla, 2012, "A Comprehensive Study of
   Nearby Early B-Type Stars and Implications for Stellar and Galactic
   Evolution and Interstellar Dust Models", *Astronomy & Astrophysics*,
   volume 539, page 143.
      Korotin, S. A. and T. A. Ryabchikova, 2018, "Non-LTE Effects of
   Helium Lines in Late-B and A Stars", *Astronomy Letters*, volume
   44, issue 10, page 621.

7 Esteban, C. et al., 2004, "A Reappraisal of the Chemical Composition
   of the Orion Nebula Based on Very Large Telescope Echelle
   Spectrophotometry", *Monthly Notices of the Royal Astronomical Society*,
   volume 355, issue 1, page 229.

8   Izotov, Y. I., G. Stasińska and N. G. Guseva, 2013, "Primordial ^{4}He
    Abundance: a Determination Based on the Largest Sample of H II
    Regions with a Methodology Tested on Model H II Regions",
    Astronomy & Astrophysics, volume 558, id. A57.

9   Poisson processes have a long history of useful applications in
    science and beyond. For example, in a celebrated early application
    of Poisson's mathematics, the nineteenth century Russian
    economist Ladislaus Bortkiewicz studied how many blacksmiths in
    the Prussian army died as a result of being kicked in the head whilst
    shoeing a horse. See, blog.minitab.com/blog/quality-data-analysis-
    and-statistics/no-horsing-around-with-the-poisson-distribution-
    troops [accessed 11 March 2019].

10 Tanabashi, M. et al. (Particle Data Group), 2018, *Review of Particle
    Physics, Physical Review D*, volume 98, id. 030001.

11 Cooke, Ryan J., Max Pettini, and Charles C. Steidel, 2018, "One
    Percent Determination of the Primordial Deuterium Abundance",
    The Astrophysical Journal, volume 855, page 102.

12 Rees, Martin, 2012, "The Limits of Science", *New Statesman*, 7 May.

13 We have borrowed this title from Marcus Chown's excellent book
    on the formation of atoms in the universe, *The Magic Furnace* (Oxford
    University Press, 2001). We recommend this and all of Marcus's
    books on science.

14 Page 68 of Rutherford, Ernest and John A. Ratcliffe, 1938, "Forty
    Years of Physics", in *Background to Modern Science*, edited by Joseph
    Needham and Walter Pagel, Cambridge University Press.

15 The atomic mass unit (or amu) is a much handier mass scale for
    discussing atoms. It is roughly the mass of a proton or neutron;
    specifically it is defined to be one twelfth of the mass of a carbon-12
    atom: 1 amu = $1.66053904 \times 10^{-27}$ kg.

16 From the 1940s to the 1970s, various militaries built and tested
    nuclear weapons, sometime accurately predicting the explosive
    power, and other times getting it quite wrong. For example, the
    Castle Bravo hydrogen bomb test was three times more explosive
    than expected. But now anyone (with appropriate security
    clearance) can simulate explosions from a new bomb design, using
    supercomputers to model the nuclear reactions and the explosive
    aftermath. The US Department of Energy is happy to tell you all

about it on their website: www.energy.gov/articles/why-nuclear-stockpile-needs-supercomputers [accessed: 10 March 2019], though there is no mention of if and how Matthew Broderick is involved.

17   The equations of big bang nucleosynthesis form a complicated network, but each part of the network is reasonably simple. For those familiar with mathematics, the equations are coupled first-order differential equations, tracking the flow of one element into another. Such equations have a wide variety of applications in physics, as well as other fields of science, finance, and engineering. See, *The Mathematics of Zombie Outbreaks* by Robert J. Smith (University of Ottawa Press, 2014).

18   Find AlterBBN online at alterbbn.hepforge.org [accessed: 11 March 2019].

19   Predictions come from: Coc, Alain, 2013, "Standard Big-Bang Nucleosynthesis after Planck", *arXiv*: arxiv.org/abs/1307.6955.

Chapter 8

1   Gross, David J., 1996, "The Role of Symmetry in Fundamental Physics", *Proceedings of the National Academy of Sciences of the United States of America*, volume 93, page 14256.

2   Augustine, *City of God*, AD 426. Note well that, like virtually every thinker since the Greeks, Augustine was aware that the Earth is spherical.

3   Cartographers have battled to represent the curved surface of the Earth on two-dimensional paper maps for much of human history. Although, as this recent story in the *Daily Mail* demonstrated, the distortions inherent in portraying a globe in two dimensions are still a shock to some. See www.dailymail.co.uk/sciencetech/article-3745016/Our-world-maps-WRONG-Countries-nearer-poles-distorted-appear-bigger-are.html [accessed: 11 March 2019].

4   Hawking, S. W. and D. N. Page, 1998, "How Probable Is Inflation?", *Nuclear Physics B*, volume 298, page 789.

Carroll, Sean M. and Heywood Tam, 2010, "Unitary Evolution and Cosmological Fine-Tuning", *arXiv*: arxiv.org/abs/1007.1417.

Evrard, Guillaume and Peter Coles, 1995, "Getting the Measure of the Flatness Problem", *Classical and Quantum Gravity*, volume 12, page L93–97.

5 Spoiler alert: in Fred Hoyle's novel *The Black Cloud* (William Heinemann Publishing, 1957), the sentient cloud tells the scientists that the universe has existed forever. The scientists "exchanged a glance as if to say: 'Oh-ho, there we go. That's one in the eye for the exploding-universe boys.'" Hoyle, an advocate of steady-state theory, obviously couldn't resist a dig at the big bang theory. We'll borrow this entertaining name for inflation advocates.

6 The video is still on YouTube: www.youtube.com/watch%3fv= ZlfIVEy_YOA [accessed: 11 March 2019]. Admittedly, we were cringing too much to watch to the end.

7 Ade, P. A. R. et al. (BICEP2 Collaboration), 2014, "Detection of B-Mode Polarization at Degree Angular Scales by BICEP2", *Physical Review Letters*, volume 112, issue 24.

8 Quoted in nautil.us/issue/59/connections/how-my-nobel-dream-bit-the-dust [accessed: 11 March 2019]. See also Keating's book *Losing the Nobel Prize*, (W. W. Norton & Company, 2018).

9 Steinhardt, Paul, 2014, "Big Bang Blunder Bursts the Multiverse Bubble", *Nature*, volume 510.

Chapter 9

1 Lewis, Geraint F., 2013, "Matter Matters: Unphysical Properties of the $R_h = ct$ Universe", *Monthly Notices of the Royal Astronomical Society*, volume 432, issue 3, page 2324.

van Oirschot, Pim, Juliana Kwan and Geraint F. Lewis, 2010, "Through the Looking Glass: Why the "Cosmic Horizon" is Not a

Horizon", *Monthly Notices of the Royal Astronomical Society*, volume 404, issue 4, page 1633.

2 Lewis, Geraint F., Luke A. Barnes and Rajesh Kaushik, 2016, "Primordial Nucleosynthesis in the $R_h = ct$ Cosmology: Pouring Cold Water on the Simmering Universe", *Monthly Notices of the Royal Astronomical Society*, volume 460, issue 1, page 291.

3 Bilicki, Maciej and Marina Seikel, 2012, "We Do Not Live in the $R_h = ct$ Universe", *Monthly Notices of the Royal Astronomical Society*, volume 425, issue 3, page 1664.

4 Karlsson, K. G., 1971, "Possible Discretization of Quasar Redshifts", *Astronomy & Astrophysics*, volume 13, page 333.

5 Repin, S. V., B. V. Komberg and V. N. Lukash, 2012, "Absence of a Periodic Component in the Quasar z Distribution", *Astronomy Reports*, volume 56, issue 9, page 702.

6 Tifft, W. G., 1995, "Redshift Quantization – A Review", *Astrophysics and Space Science*, volume 227, pages 25–39.

7 Other researchers, including Halton Arp, claimed that quasar-galaxy pairs showed redshift quantization. This is on small scales, and so different from the Karlsson claim. But again, if correct, it would only rewrite quasar astrophysics not the big bang theory. And the data is not particularly convincing when we use larger data sets; see E. Hawkins, S. J. Maddox, and M. R. Merrifield, 2002, "No Periodicities in 2dF Redshift Survey Data", *Monthly Notices of the Royal Astronomical Society*, volume 336, issue 1.

8 Nordgren, T. E., Y. Terzian, and E. E. Salpeter, 1996, "The Distribution of Galaxy Pair Redshifts", *Astrophysics and Space Science*, volume 244, issue 1–2, page 65.

9 Broadhurst, T. J., R. S. Ellis, D. C. Koo, and A. S. Szalay, 1990, "Large-Scale Distribution of Galaxies at the Galactic Poles", *Nature*, volume 343, page 726.

10 Einasto, J. et al., 1997, "A 120-Mpc Periodicity in the Three-Dimensional Distribution of Galaxy Superclusters", *Nature*, volume 385, issue 6612, page 139.

11 Kaiser, N. and J. A. Peacock, 1991, "Power-Spectrum Analysis of One-Dimensional Redshift Surveys", *Astrophysical Journal*, volume 379, page 482.

12 Yoshida, N. et al., 2001, "Simulations of Deep Pencil-Beam Redshift Surveys", *Monthly Notices of the Royal Astronomical Society*, volume 325, issue 2, page 803.

13 Smoller, Joel and Blake Temple, 2003, "Shock-Wave Cosmology inside a Black Hole", *Proceedings of the National Academy of Sciences*, volume 100, page 11216.

14 Among Christian and Jewish scholars (and believers), opinion is divided over the interpretation of the first chapters of Genesis. A wide range of interpretations have been offered, stretching back millennia. See, for example, Collins, C. John *Reading Genesis Well: Navigating History, Poetry, Science, and Truth in Genesis 1-11* (Zondervan Academic, 2018).

15 Source: Chandrasekhar, S., 1987, *Truth and Beauty: Aesthetics and Motivations in Science*, University of Chicago Press (page 20).

16 Lewis, C. S., 1944, "Bulverism", republished in *God in the Dock* (Eerdmans 1970).

17 Conner, Samuel R., and Don N. Page, 1998, "Starlight and Time Is the Big Bang", *CEN Technical Journal*, volume 12, issue 2.

  Ross, H., 2000, "Humphreys' New Vistas of Space", *CEN Technical Journal*, volume 13, issue 1.

  Fackerell, E. D. and C. B. G. McIntosh, 2000, "Errors in Humphreys' Cosmological Model", *CEN Technical Journal*, volume 14, issue 2.

18 Humphreys, D. Russell, 1998, "New Vistas of Space-Time Rebut the Critics", *CEN Technical Journal*, volume 12, issue 2.

19 Hartnett, John G., 2003, "Look-Back Time in Our Galactic Neighbourhood Leads to a New Cosmogony", *Technical Journal*, volume 17, issue 1.

20 See Robert A. Stokes's contribution (page 323) to *Finding the Big Bang*, edited by P. James, E. Peebles, Lyman A. Page Jr., and R. Bruce Partridge (Cambridge University Press, 2009).

21 Measurements from pre-1997 are helpfully compiled in Smoot, George, 1997, "The CMB Spectrum", preprint: arxiv.org/abs/astro-ph/9705101v2, while more recent measurements can be found on NASA's LAMBDA database: lambda .gsfc.nasa.gov

Chapter 10

1 Lynden-Bell, D. ,1976, "Dwarf Galaxies and Globular Clusters in High Velocity Hydrogen Streams", *Monthly Notices of the Royal Astronomical Society*, volume 174, pages 695–710.

2 Kroupa, P., C. Theis, and C. M. Boily, 2005, "The Great Disk of Milky-Way Satellites and Cosmological Sub-structures", *Astronomy and Astrophysics*, volume 431, pages 517–521.

3 Wheeler, John, 1997, *At Home in the Universe*, American Institute of Physics.

FURTHER READING

Popular Level Introductions to Cosmology

Barrow, John, 2012, *The Book of Universes*, W. W. Norton & Company.

Chown, Marcus, 2001, *The Magic Furnace*, Oxford University Press.

Davies, Paul, 1997, *The Last Three Minutes*, Basic Books.

Ferris, Timothy, 1998, *The Whole Shebang*, Simon & Schuster.

Greene, Brian, 2005, *The Fabric of the Cosmos*, Vintage.

Harrison, Edward, 2000, *Cosmology: The Science of the Universe*, Cambridge University Press. [Highly recommended]

Hawking, Stephen, 1988, *A Brief History of Time*, Bantam.

Natarajan, Priyamvada, 2016, *Mapping the Heavens: The Radical Scientific Ideas That Reveal the Cosmos*, Yale University Press.

Rees, Martin, 2003, *Our Cosmic Habitat*, Princeton University Press.

Weinberg, Steven, 1993, *The First Three Minutes*, Basic Books.

Vilenkin, Alex, 2007, *Many Worlds in One*, Hill and Wang.

In addition, anything by Sean Carroll, John Gribbin, Janna Levin, Alan Lightman, Lisa Randall, Carlo Rovelli, and Leonard Susskind is worth a read.

Cosmology Textbooks

Carroll, Bradley W. and Dale A. Ostlie, 2017, *An Introduction to Modern Astrophysics*, Cambridge University Press. [Beginner]

Dodelson, Scott, 2003, *Modern Cosmology*, Academic Press. [Advanced]

Liddle, Andrew, 2003, *An Introduction to Modern Cosmology*, Wiley-Blackwell. [Beginner]

Mo, Houjun, Frank van den Bosch, and Simon White, 2010, *Galaxy Formation and Evolution*, Cambridge University Press. [Intermediate]

Mukhanov, Viatcheslav, 2005, *Physical Foundations of Cosmology*, Cambridge University Press. [Advanced]

Peacock, John, 1999, *Cosmological Physics*, Cambridge University Press. [Advanced]

Ryden, Barbara, 2016, *Introduction to Cosmology*, Cambridge University Press. [Beginner]

INDEX